Ends.

Why we overlook endings for humans, products, services and digital. And why we shouldn't.

Joe Macleod

ISBN 13: 978-9163936449
ISBN 10: 9163936445

Contents

5   **About the author**
7   **Acknowledgements**
9   **Preface**
15   **Introduction**
19   **The biased customer lifecycle**
31   **The social context of closure**
49   **Introduction to chapters 3 & 4**
51   **The distancing and the fading**
69   **The quickening and the tethering**
87   **The psychology of endings**
105   **The narrative of endings**
119   **Love and marriage...for life?**
131   **An intangible tale of services**
147   **Saying goodbye to your products**
167   **Closure in digital**
193   **Business oversight**
215   **Conclusion.**
227   **Ends. notes**
239   **Index**

Obligatory author picture

About the author

Joe Macleod has been working on the issue of appropriate endings and closure experiences for fifteen years. Through his work in design, technology and services, he has detected a common pattern of denial at the end of the customer lifecycle. In the last couple of years this interest has led him to establish a research project based on sharing this insight and new approach with people via conferences, articles, teaching, projects and now a book.

His 20-year professional career has been in web, telecoms and carrier companies, where he led teams and built a variety of successful products. Most recently as Head of Design at the award-winning digital product studio Ustwo, he built a globally recognised team, working with the world's favourite brands on the most pioneering of products.

A regular speaker and commentator of design and design education as well as the founder of the IncludeDesign campaign that brought the UK's leading designers together to defend creative education.

He now works on the Closure Experiences project; researching, talking, consulting and writing about this important yet overlooked issue.

Ends

Acknowledgements

This story couldn't have happened without the encouragement of Nigel Shardlow, a friend whom I worked with long time ago. He was an early reader of my work who believed in the idea and helped to frame the early concepts.

My work developed and matured, the book started to take form. It would have been difficult to progress this without professional help (especially as I am pretty dyslexic). I would like to thank Kate Goldstone who did early edits and corrections.

Also, I would like to thank Monica Shelley, who has been fantastic as a teacher, editor, contributor and wise sage on consumer affairs. She also happens to be my mother in law.

I would like to thank Danielle Newnham for her inspiring war stories around publishing, and the encouragement to pursue the route of going it alone with self publishing.

Two more people I'd like to thank are Steve Bitten, who I worked with for many years and has been a great help in getting the message out to an audience and Alvaro Arregui, who has been wonderful in providing a vision for the look of the book.

Not to be forgotten are Mills and Sinx of ustwo, and their

Ends.

enormous encouragement in all matters. I have achieved a great deal in the past, infected by their enthusiasm, and the pursuit of this dream was also in some part down to their encouragement.

I would like to mention a dear friend, Chris Downs without whose support, mentoring and inspiration I'd be lost. He has been in the crowd at the events, cheering from the back and an ever-present person in my life.

Finally, I wouldn't have got anywhere at all without my guide and wife, Alexis, who holds all the family together through her love, encouragement, warmth and happiness.

Preface

Around the early noughties, I had a couple of experiences that would profoundly change my opinion about consumerism and ignite the investigation that I am about to share with you.

The first was using a new service from the mobile phone company Orange. It was a voice recognition avatar/secretary on my phone called Wildfire. It promised to be a pioneering achievement of natural language interface of the time. Orange believed it was going to revolutionise the way we related to technology. I signed up with excitement, believing that I was about to enter into a magical world of pleasant and convenient conversations with an automated assistant.

As with all great technology claims, this one soon faded away when confronted with reality. This particular reality was brought into sharp focus near roads, with heavy traffic. Wildfire couldn't hear me. The shouting match usually went something like...

Joe: "Wildfire. Play answerphone messages."
Wildfire: "Sorry, I don't understand you."
Joe: (louder) "Wildfire. Play answerphone messages."
Wildfire: "Sorry, I don't understand you."
Joe: (LOUDER) "Wildfire! Play answerphone messages." (Insert curses and rants)

Ends.

I got to the point where I hated the Wildfire avatar so much, that to terminate the service using the normally emotionless method of simply cancelling it wouldn't satisfy my anger and frustration. I wanted something more powerful, something more appropriate that matched up to the dreams I had when I responded to the sales pitch and signed up to the service. I wanted to strangle Wildfire until its little avatar eyes flickered, then faded to a dead cold black. I wanted emotional satisfaction in my service ending.

The second experience was teaching students on the graphic design course at Central St Martins in London. I set them a brief around consumer waste for them to practise their skills - it's something that's probably been done hundreds of times before in schools and colleges around the world. After gathering into groups they went off to consider the issue, do some research and create their design proposals. Eventually they returned to share their ideas and describe their designs.

Student after student presented their solutions and proposals - but as they did so, they described more and more objects and paraphernalia. Instead of responding to the brief and demonstrating their awareness of the implications of creating more and more waste, they simply added to it. So, although they had responded technically to the brief, they had totally failed to understand the philosophy involved.

These two experiences left me with the clear impression that we don't have a vocabulary around endings. We have only the language of make more stuff and create more new things.

I grappled with this issue over the years, reflecting on it over my early career and assessing it against the work I was doing. Time and time again I saw this pattern of denial about endings repeat itself. The off-boarding of the customer was being ignored, as was the long term impact of his or her consumption.

Soon after my initial realisation, I had the chance to investigate the issue through a project for one of my clients. This provided the opportunity to answer some basic questions and set down early ideas.

The project was well received, one person even saying it was one of the most important pieces of work they had seen.

Although this was reassuring, it was impossible to pursue the issue full time and carry on working for a living. So it became a hobby for many years. When I had read or seen something relevant, then I recorded my reactions and thoughts in sketchbooks, writing down notes and ideas I had around this issue.

Over this time I also saw evidence of the issue through the roles I had in various sectors of the design and tech industries. All of this evidence provided examples of goods and services we sold to customers - all were promoted with wonderful, inspiring starts to the purchase process, yet failed to acknowledge to the customer that these goods and services actually had a lifespan.

At the zenith of the dot.com boom there was total, blinding confidence in technology. Everyone believed that what was being created would last forever! Dubious business models were employed that aspired to change the world, yet ignored the potential lifespan of our creations.

It was great business until it wasn't and then it really was awful. Work dried up. Panic set in. And we were making people redundant wherever we could.

In the product designer culture of some of the big multi-nationals where I worked, design teams of hundreds of people had enormous budgets to research every detail of all kinds of people, from all types of background, country and culture. But absolutely no attention was paid to what happened to the products when the consumers no longer used them. As a result, millions of phones are now left in drawers, with their owners fearful of what might happen to the data on them and what damage the old phones might do in landfill. They lack the language of endings.

Working in the emergent industry of service design, I was privileged (through some of the leading lights in the industry) to see how the service design approached this issue. Once again we confidently created services that people only ever signed up to, and started. We rarely designed off-boarding or endings.

More recently I found myself back in the digital industry, and this time the business was more mature and responsible than in the wild west of dot.com days. But we still drove businesses to create services that only led people in. How they left these services was never considered. What happened to their data was overlooked. Once again, we were in denial that things came to an end.

Throughout this time the concept of closure experiences was on my mind. Every now and then another piece of the puzzle fell into place and another aspect of culture aligned with my thinking. All the time we were all creating more and more products and services which totally lacked appropriate life cycle endings.

In 2015 my wife and I decided to change our lives around. She wanted to get her career back on track after 9 years of raising our kids, I wanted to spend more time with our two growing boys. I was also keen to deal with this issue - closure experiences - which had been weighing on my mind

The first thing I needed to do was figure out a plan and my goals. Initially, these were fairly basic: research the issue, write about my findings and share this online or through conferences and workshops. I expected this to last about 6 months, and felt I would have exhausted my interest in it by then. However, this is far from being the case, as I have dug into this issue, it has revealed itself to be something far bigger, far more important, and far more complex than I could ever have imagined.

Our relationship with endings has a long and complex history, being shaped by enormous changes in society, from the transition from nomadic wandering to more static societies, to changes in the concept of possession and ownership, to what we imagine heaven to be like, and the significance of funerals. These have influenced our relationship with endings, closure and consumption in Western society today.

I found it immensely intriguing to research the elements and motivations that brought about this bias in our relationship with products and services. I wanted to dig more and more in history and share this fascinating story. What was originally a project to share some thoughts in talks and blog posts became this evangelical mission to get the message out there and tell this story, and change the way we deal with the life history of goods and products.

The Ars moriendi ("The Art of Dying") are two related Latin texts dating from about 1415 and 1450 which offer advice on the protocols and procedures of a good death, explaining how to "die well" according to Christian precepts of the late Middle Ages. Wikimedia Commons

Introduction

"*'Tis impossible to be sure of anything but Death and Taxes,*"
Christopher Bullock, The Cobler of Preston (1716)

Taxes? Yes. But death? This may have been the case once, but is it
so now? In recent years we have changed the whole approach to the
process and permanency of death with the introduction of technology
such as breathing machines, complex definitions of brain death and
the potential to deep freeze terminal patients in the hope that one day -
when more cures are available - they can be re-woken to life.

Emotionally, death and the concept of heaven have lost their
meaning for some people as we progress towards a more secular
society. And, with the introduction of hospitals and hospices, we are
losing the personal experience of death and dying, since few of us are
able to be present at the last hours of our loved ones.

This distancing of death has been mimicked in our consumers
experiences. As we ignore the (metaphorical) death of the products
and services we buy. In contrast, other human made experiences, such
as films, books, games, have a clear start, middle and an end. These
stimulate thought and meaning to the viewer, reader or player. But
in consumer experiences, after we have used a product, exhausted
a service, or given up on the latest app, we fail to acknowledge that

ending. This means that an important tool for dealing with reflection, meaning and responsibility around consumption has been removed. I would argue that this has fuelled the failures linked to consumption.

The statement 'Too big to fail' justified saving the worst of banks in the 2008 crash from their self inflicted endings. Denying that this was the case more and more money was borrowed, and the status quo was propped up, despite the logic of established economics. 'Right to be Forgotten', is the ambitious law of the European Union that protects a person's rights in a digital world which protects itself from loss and refuses to allow us to remove those items that we were persuaded so convincingly to share. There have been nearly 30 years of discussion and debate about climate change - and still we refuse to accept the implications of failing to cut carbon emissions. Long ago we lost the necessary vocabulary, the skills we need to make those vital changes. Once upon a time we had, in contrast to the present day, a rich vocabulary of endings which supported our culture of reflection and responsibility.

In its place we evolved a biased consumer lifecycle that has limited our vocabulary to the description of the new, avoiding the need to discuss death and endings.

This book tells the story of valuable lost endings. It charts the history of endings in our lives and our consuming habits. It argues that a social change in our relationship with endings is responsible for society's biggest problems. It champions re-engagement with these valuable endings, with increased attention to designing, developing and highlighting them in our consumer experiences.

The story in this book goes back to the early stage in human history when we stopped being nomads, filled our dwellings with goods and products, buried our loved ones with ceremony and goods they might need in the afterlife. It then moves on to the industrial revolution, the birth of commerce and the growth of consumption. The moon landing, the awareness of germs and the atom bomb are shown to contribute to this story. It introduces parallel paths of inspiration that framed our consumer lives, how the mind works, how it relates to death and how it creates meaning in endings. It examples work in the film industry and the way we tell tales in narrative and games.

It carries the argument for improving endings into the current consumer landscape with reflections around the consumption of physical products, the services industry, and the emerging digital industry. It provides specific examples, holding some up as good practice, and others as bad.

The story I tell couldn't possibly cover all aspects of the argument I am making. On the subject of human death alone there is a lifetime's worth of research. Instead it touches on many issues to lay the seed of an augment and open the reader's eyes to alternative approaches to consumption.

Some of the readers of this book might be creators of products, services or digital products, which is a background I have myself. For these people I hope this book inspires you to approach your creations with a different strategy which challenges the accepted and established thought of businesses and environmentalism. With any luck it will empower you to take action positively armed with the insights and arguments spelt out in the chapters that follow.

We are all consumers. This book can aid reflection upon our consumer behaviour in deeper and more profound ways. It aims to frame our behaviour by locating it in the context of a society which has been changing under the influence of centuries of consumerism. I hope this book inspires you to reflect about your own consumption and to see the bias in it.

But most of all I hope that you find this book as interesting to read as I have found it inspiring to write.

Chapter 1

The biased customer lifecycle

We are all consumers. We buy and use products and services, each of which has a life cycle for us spanning from purchase to useless. The way we consume has come a long way from the earliest beginnings, when consumption consisted of a crude transaction or debt: for example, exchanging the corn harvested in autumn for furs to keep the family warm in winter.

Nowadays we buy things on a whim, impulsively, to satisfy an emotional desire which has been constructed by influences embedded in a complicated social hierarchy. This hierarchy has evolved alongside the basic features of modern life such as money, markets, advertising, and its sole aim is to target and inspire consumers to buy the goods that have been created.

This evolution has created many jobs, a great deal of wealth and, you could argue, a certain purpose for the billions of people that now populate this planet. But it has its downside: occasionally and often in the long term, it fails.

The physical aftermath of our consumption of products can be seen in the problems of pollution. Plastic litters the oceans, we discuss the warming of the planet while ice melts. These are the bad outcomes from our consumption, the aftermath that we fail to capture before it leaks into the natural environment around us.

In the world of services the goods supplied are less tangible. They could be the healthcare you receive from a nurse, the plane ride you buy from a travel company or the money management carried out by your bank. Instead of physical waste, here the bad outcomes are less tangible: mis-selling in banking, lost pensions and locked-in gym membership.

Even less tangible are the bad outcomes from the new landscape of the world of digital. You can see examples in the software you have on your computer, the apps you download to your phone and the social media applications you use. The problems arising from this industry often get ignored as they are invisible, hidden on servers, and in your data that companies keep and use in ways that may well cause you anxiety in your daily life.

I see the problems that arise from these different areas of consumption as having the same source. Endings are no longer part of the overall consumer experience. We have moved the source of the problem away from the cause.

As consumers we are able to overlook endings. In business we have built a culture of ignoring them. As students we are taught they are not important. Endings are dodged and left for someone else to clear up. They are broken away from the rest of the experience.

It would be simplistic to assign blame. This is in fact a deep societal problem which has developed over hundreds of years. We have been programmed to ignore endings, to consider only the short term and to deny that endings need to happen.

We need to understand the position we now find ourselves in. It's not enough to design better endings, although that would be an excellent start. It is about understanding the full significance of the position we are in: emotionally, environmentally, socially and commercially.

To do this is no small task, but I have attempted to put together the arguments, evidence and potential solutions for what I believe defines this enormous problem with endings.

Some background and models.

First of all, let's look at some models, terms and systems that will help to frame the discussion spelt out in the rest of the book.

The consumer lifecycle

The consumer lifecycle is a model used in marketing to define stages through which a person goes when purchasing and consuming something. For example, as a consumer you recognise you have a need to buy something. Maybe it's more milk, or something more

need   consideration   selection / sign-up   first time use   continued use   loyalty   renewal /repair   delete / dispose

The consumer lifecycle

complicated like a new car, or to sign up to an Internet provider. Then you might look at some different options for where you are going to get this new purchase. Once you have decided, you then make the transaction, hand over the money, sign up and commit to the service. You then use the service or product until a point when it is not required. The science of this process can be reflected as a consumer lifecycle.

Starting experiences and closure experiences

We will build on the consumer lifecycle and adopt a common vocabulary for the discussion in the book. To do this we'll simplify the product or service lifecycle into three phases:

on-boarding, usage, and off-boarding.

In this context, on-boarding is made up of starting experiences. These are the actions that persuade the customer to commit to the product or service: they form the start of the relationship between consumer and provider. Examples of starting experiences are advertising, that attracts you to a product or service, marketing that orientates you towards the decision to purchase, packaging that enhances the product and makes it look more attractive and Terms and Conditions that establish the agreement between provider and customer.

The usage phase completes tasks, empowers people and orders chaos. It is the stable committed relationship between the consumer and the service they use, or the product they own. Examples of usage

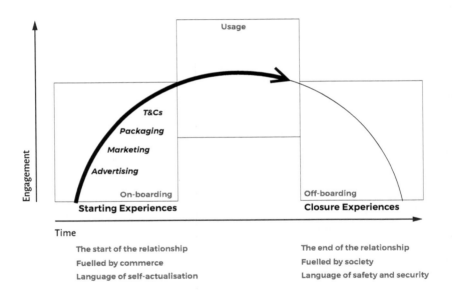

experiences are paying into or drawing from a pension scheme, daily usage of a car or regular usage of an app.

Off-boarding is a less well-acknowledged part of the product or service lifecycle. It is made up of closure experiences of different kinds, such as the effort needed to neutralise the effect of the closure, to terminate the relationship between consumer and purchase. Typical closure experiences are the completion of a mortgage, the deletion of unwanted photos online, the appropriate disposal of an unwanted product, saying goodbye or closing unused accounts. Off-boarding acts to tidy up the impacts of consumption, and neutralise it's ills.

The consumer awareness of endings

Sadly, most of industry ignores the ending as an important part of consumption. Although many producers make great efforts to counter the negative effects of consumption at an industrial level through legislation, improved materials and corporate governance, they are terrified of sharing, or even acknowledging endings with the consumer. Both producers and consumers have learnt to deny that endings happen.

Consequently we have all lost control of this part of the consumer relationship, which in turn removes our ability to deal with it sensibly. The product or service life cycle that has been created protects the consumer from questioning the consequences of consumption. A myth has been created for consumers that it is better to re-purchase rather than end the life cycle in a responsible manner.

By failing to acknowledge the importance of dealing with endings we lose our ability to improve them. The beginning, or on-boarding, of a customer is now a complex ritual that, as consumers, we navigate confidently. But the ending of that process has been distanced from the consumer and is now almost invisible as part of the same experience. That distancing has also removed the consumer's individual responsibility. This has had critical knock-on effects to the way we relate to the products and services that we consume, and is now haunting our ability to deal with personal and global issues.

Products

The product industry has a great deal of experience in the issue of problematic endings. Hundreds of years ago the process for purchasing and owning a product was far different. Many of the products we purchased would have been produced locally, maybe by someone we knew in the same village or town. We would have had previous knowledge of the producer and there would probably have been a measure of personal trust between us, more particularly as they would have a reputation to uphold in the community.

Once purchased, we would have a responsibility to the object. We would have valued it differently and mended it if it broke. We would also have taken responsibility over its disposal when its use came to an end.

Although there was a great deal of dumping of waste, there were no organised rubbish dumps that processed waste. We would have had to burn what we could, then live with what we couldn't. The end of that product relationship required an acknowledgement, reflection and responsibility as we were living with the consequences.

Our more recent product relationships are far different. Industrialisation has meant that access to products has became more frequent, less personal. The relationships that were established between provider and purchaser in small communities have long since gone to be replaced with 'brands' that serve as a proxy of trust in the absence of a person as a producer. The varieties of choice have become enormous with 20,000 unique consumer packaged goods being launched every month worldwide. That's 650 a day, by the way.[1]

To make way for more purchasing we have become efficient at removing our individual responsibilities at the end of the product relationship. We throw unwanted items in the trash, which are then conveniently removed from our lives by the municipal companies acting for local government and taken to distant and out of sight landfills. This removes the consequences for which we would previously have had to bear the responsibility and frees us to consume yet another product.

Services

The service industry has reflected the product industry in its distancing of responsibility at the end of a customer lifecycle. There are many examples that I will go into later in the book, but I will choose pensions in this instance.

We have increased the number of employers we have had over our careers - currently 11 on average in the UK, according to the Department of Work and Pensions.[2] Each one provides us with a pension pot to invest in. Sadly, the charity Age Concern, found that 1 in 4[3] of these pensions goes missing because of difficulties of connection over a lifetime between provider and pensioner. This is an inevitable outcome of the delivery of a service over decades as people move house and jobs, companies merge and change names - basically, life passes by and our consumer culture is too busy to wait around to keep in touch.

The financial services industry, amongst many, has been questioned over its short term approach to pensions and many other financial products. But this is only half the story. Industry has been permitted to develop this denial of endings as a result of our own consumer desires for the new. The real problem is both consumer and provider have developed a denial of endings, so that we have lost the vocabulary to discuss, recognise and resolve the consequent dilemmas.

Digital

Our social experiences have shifted from being emotional to becoming distant, systematic and open-ended. This ignores the need for long term reputation and reliability and focuses on a desire for short term 'likes'.

In previous generations we might have sat down with friends and family to share a photo. I am sure many of us remember doing this with our grandparents decades ago. We slowly turned pages, and they explained who was in the picture. They would supplement the photo with anecdotes about the person, place and times gone by. It's hard to imagine this event having many negative knock-on consequences. It was a private showing, and not recorded. It was a simple, long established act of passing down knowledge through stories and images to loved ones.

Changes in recent years, with the explosion of social media, have enabled us to share photos with friends, families and in fact anyone who might be interested. Despite the enormous benefits this has brought in speed, breadth of distribution and establishing identity, it has also raised questions around legacy, reputation and privacy. The knock-on consequences might be far more impactful than we can imagine in the moment of taking and sharing a picture.

Courtesy of Tim Bennett @flashman

This photo captures poignantly the issues arising from the lack of control and conclusion we have in our modern photos - although most of us probably won't be unknowingly exposing the location of a rhino. It is more likely that we will be undermining ourselves with lingering inappropriate pictures from parties, comments made in the heat of a moment, or even pictures of our children in their early years, that they don't appreciate when they are older.

The point is not how we are going to deal with pictures lingering online, the problem is the creation of systems that can remove these images once they have outlived their usefulness. Consumers should have just as much control at the end of the relationship as they do at the beginning. We have created infinite systems of sharing online, but precious few methods of removal. We need to rebalance the opportunities to remove with the infinite opportunities to share. What we need is a balance between starting and ending experiences.

The emotional imbalance

Our motivations and emotions at the beginning and the ending of the customer lifecycle vary. Not only do the tone of language and the source of the messages change, the emotions and motivations are quite different.

The examples above indicate that the sources of the messages associated with a closure experience often differ from those that coax the consumer to buy or join at the beginning of the product or service lifecycle.

The tone of off-boarding is also starkly different - an austerer, emotionless tone, desperately trying to demonstrate the impartial voice of authority in the black on white text that is often used. Whereas the voice of on-boarding is encouraging, full of joy, inviting our normal life to get better through colourful language, imagery and messages.

These messages speak to different motivations. The off-boarding asking for restraint, consideration and champions safety and security approaches, with a worrying tone of foreboding.

The on-boarding encourages potential, tickling your dreams and fantasies with suggestions of how you can fulfil them and reach a level of self-actualisation.

We can illustrate this with the well used model of Maslow's

Ends.

Maslow's hierarchy of needs. 1943

Hierarchy of Needs [4]. Starting Experiences that facilitate the on-boarding of a customer answer, or at least propose to answer some of our strongest motivations - are represented by the arrow on the left. In contrast, Closure Experiences, that aid the off-boarding of a customer seek to spell out a far humbler set of motivations - indicated by the arrow on the right.

Differing voices

It's very easy to spot the way in which the product lifecycle is biased by different messages by simply looking around you every day.

At the on-boarding phase of the product or service lifecycle we hear wonderful stories of how our lives could be improved by the purchase of a new product or service. But by the time of off-boarding we are made aware of the impact of our purchase and the damage this has on the wider environment.

The parties involved have different responsibilities and perceptions. At one end we have producers and advertisers that want to sell a product. Their responsibility is to the shareholders of the company to create a good turnover. At the other end we have governments and agencies who are responsible for safety and good practice on behalf of society.

Our empowerment is encouraged at one end of the process and derided at the other. For the consumer this is conflicting and overwhelming.

I will refer to these issues and models in the rest of the book as we analyse particular models to highlight the broken model underneath. We will look at examples from industry and how starting and ending experiences are used. We will also consider how these are communicated to customers throughout the consumer lifecycle.

We will start by considering the long history of endings.

Michel Serre, Scène de la peste de 1720 à la Tourette (Marseille), Wiki Commons

Chapter 2

The social context of closure

Our ancient ancestors experienced nothing of the abundance we do today. In their tough daily existence they couldn't have imagined the future bounty that humans in the wealthy West would experience thousands of years later. Their close relationship with death, and their precarious relationship with life, left little space for creative interpretation of any kind.

Back then, when people died, it was a simple. Closure was a clear function of life. There is some evidence that Neanderthal people buried their dead intentionally with some ceremony, but experts doubt it was anything more than bodies simply being disposed of for secular or other practical reasons. The earliest of deliberate burials really didn't happen until the nomads came along.

Our nomadic ancestors also had short and difficult lives. Their travelling lifestyle required simple tools and restricted them to just a few possessions. Mobility was the focus of their existence. Being able to travel quickly to safety or find new grazing and foraging areas was their priority. This meant groups faced difficult decisions when an individual became a burden to others - examples have been found of disabled newborns, the elderly or ill being left as the group moved on to new places. Death was crude and common, as were the shallow graves lacking in ceremony, indicating the impracticality of dressing death up and creating a fanfare when your core priority was to move on to the next grassland and survive another day.

Finally, in the Middle Palaeolithic period (300,000-50,000), we start to see more meaning given to the closure of life. Burials show symbolic intention and individuals seem to have been buried with grave goods, suggesting belief in an afterlife when these items would be needed by the dead person. As Philip Lieberman, a cognitive scientist and an authority on evolution suggests, it may signify a *"concern for the dead that transcends daily life"*, a belief in some kind of afterlife.

Some graves dating back as far as 60,000 BC reveal bodies in the foetal position[2], surrounded by flowers, ivory ornaments, shells and painted bones. Although there is some discussion around the purpose of the statements being made – either to symbolise re-birth or prevent the dead from rising to disturb the living - it is clear the belief system at the time included some form of life after death, a post-mortem existence, revealing a new understanding of our own mortality. They indicate people practised/carried out religious ceremonies at the end of a human life, a formal ending of sorts, an acknowledgement that something should take place to ease the transition to the next place, beyond the grave. In other words, the off-boarding of a life.

From hoarding to graveyards

The melting of the ice packs and the slow emergence of green pastures enabled nomads to stop wandering, settle down and adapt to farming, something that spread steadily from the Middle East. As permanent settlements emerged, the relationship with a formal form of closure matured alongside increased wealth, ownership and consumption. The earliest example of these maturing communities comes from the Natufians living around Syria, Jordan and Israel in about 15,000 - 11,000 bc.[3] They advanced farming methods by growing and storing wild wheat and barley, using stone sickles to harvest cereal grasses. This in turn brought about more abundance, more wealth, and ever-increasing levels of ownership. Land now became owned, or at least occupied by an individual. This in turn enhanced the concept of a person's identity since they were located in one place. Having somewhere to store possessions meant that people could accumulate more items, and even hoard them.

Being stationary on their own land meant people were better able to tend to the sick and dying. And for those who didn't get better, death could be given more meaning through a permanent grave, close to the rest of the community. Thus more and more emphasis was put on graves and their construction, which soon started to mirror a newly-emerging class system. Some members of the community had more privileged burials, revealed through the adorning of complex head-dresses, beautiful bracelets and precious necklaces. Other cultures removed the heads of the deceased, separating them from the body and covering them with plaster, after which the heads would be kept in the family home, something which would later become common in Neolithic burials. It isn't dissimilar to keeping the ashes of a departed family member on the mantelpiece today. Some cultures, for example the Babylonians, kept the remains of their closest dead family members in the same house as the living[4], just on a different floor. The graves of the dead finally had a fixed and marked location, a place that could be attended regularly by family and loved ones.

These critical steps in our cultural evolution mark the first meaningful processes, the first time the human race engaged fully with the ending of life. Respecting someone's death with rituals, objects and physical locations to reveal status and identity, and providing a memento of a life passing plus a new imaginary life being prepared for, became universal, powerful cornerstones in establishing the etiquette of endings.

The stellar rise of Heaven

The relentless grind of daily life, work, money, food and health, was a stark contrast to the endless bounty offered in Heaven. Heaven provided the carrot for how to get there. Religion provided the guidance. And a hard life for most people provided a clear incentive. Heaven is a powerful theme throughout cultural history. Many desperate people have been comforted by the potential of a Heaven, a potent currency with which to trade emotions. It incentivises the end of the experience of life.

The up-selling of death as a good way out of a hard life is an attractive option, presenting something better than mere death. To this day an afterlife is constantly over-sold to the living. The concept of a Heaven has become key to religious story-making across multiple cultures, a way to carry on life when the body expires, and the narrative is surprisingly similar across cultures. Heaven is often portrayed as a distant place, above the earth, frequently in the stars, but sometimes in the mountains. It tends to be a joyful and indulgent place where the everyday challenges of living no longer apply.

Early versions of Heaven meant homes for the Gods. Mere mortals didn't get a look-in, a situation that wasn't sustainable because it didn't attract the masses. The Egyptians philosophized about the nature of the afterlife a great deal more than many cultures before them. The historian Alan Segal believes that they were the first to associate Heaven and immortality, many millennia before ancient Greek[5] texts portrayed such a thing, and long before the Bible. Egyptian death was very even handed, with the poorest and richest being judged alike by the God Osiris. Anyone, regardless of status, could be promoted to Heaven or - alternatively - cast adrift to be consumed in a second death by Attim, a terrifying god who was part hippopotamus and part crocodile. Attempts at improving your odds in this situation are given in the Book of Data. This provided the dying Egyptian with prayers to encourage a positive judgement. The first of many such books that have since promised life-hacks to tackle death and improve the odds of a positive religious judgement. It could be argued that most religious teachings either encourage worshippers to lead a good life in preparation of a good death, give instructions about how to have a good death, or enjoy the rewards of Heaven as a result of the first two. In the modern commercial world this type of improved 'other' life, is used to create a carrot to another type of life. Many advertisements for pensions, for example, present the cultural imagery of Heaven - people dressed in white, relaxing, drinking and eating. Sadly the options of living forever in this type of pension Heaven is unachievable as a business model.

Max Weber, the 19th century sociologist defined four types of reward offered to worshippers of all religions. The first being the promise of compensation in this world, the second the immortal, spiritual elements of life. The third is the physical side of life and the fourth the doctrine of Karma, involving the sum of actions throughout one cycle of reincarnations, where good deeds lead to a better life next time around.[6] As Weber points out, not all religions present the same Heavenly format but they all come with the notion that there's another place, beyond this life, that represents an improvement on this one. And it's a pretty attractive place, designed to ease the transition from life and drive us to act in certain ways while alive.

For Buddhists the achievement of Nirvana is key, by means of extinguishing all desires and letting go of everything. Buddha believed that desire is the 'flame that burns us'[7] and keeps us attached to death and rebirth. Extinguishing that flame through meditation, for example, allows the individual to stop suffering and achieve Nirvana. Buddha is never presented as a bountiful consumer. You won't see him eating and drinking, enjoying the spoils of Heaven as suggested by other religions. Buddhist monks have relied on alms-food, donated by their fellow humans, for more than 2500 years.[8] They don't subscribe to the usual round of endless consumption as a form of Heaven.

Relatively recent religions like Jehovah Witnesses and Mormons believe in a more structured afterlife. Jehovah Witnesses see Heaven as a place with a fixed population of 144,000 people,[9] chosen ones who will ascend to Heaven and rule with God in a state of perfect health and everlasting life. If you happen to fall outside that magical 144,000 there's still hope – you'll be resurrected into some sort of Earthly Paradise when the final day of reckoning comes. Mormons have three Heavens, *"three glories of resurrected bodies: one like the sun (celestial), another as the moon (terrestrial), and the third as the stars"*[10], a sort of Gold, Silver and Bronze system. All of these Heavens come with their own comforts and pleasures, and *"...even the lowest kingdom surpasses the glory of the earth."*[11]

In the Islamic faith, Heaven is called Paradise. To get into Paradise your good actions in life have to outweigh the bad. [12]According to the Quran, *"Eat and drink at ease for that which you have sent forth (good deeds) in days past!"* (Quran 69:24). Islam's Paradise portrays many of earth's comforts and human desires. *"... there will be there all that the souls could desire, all that the eyes could delight in ..."* (Quran 43:71) *"...they will be adorned therein with bracelets of gold, and they will wear green garments of fine silk and heavy brocade. They will recline therein on raised thrones. How good [is] the recompense! How beautiful a couch [is there] to recline on!"* (Quran 18:31).[13]

According to the Bible, the Christian version of Heaven describes a walled city with twelve gates around it. The city is decorated with precious stones and gold. [14]According to the Bible the kings of the earth will bring their splendour to it, and it will be free from impure, deceitful or shameful people. As Revelation 21:4 says, *"God shall wipe away all tears from their eyes; and there shall be no more death, neither sorrow, nor crying, neither shall there be any more pain: for the former things are passed away."*[15] Get it wrong, though, and the Christian religion plunges you into a terrifying hell populated with demons, fire, flames and perpetual misery.

Hindus believe in reincarnation. The form and quality of the next life depends on a person's actions in this world. If they carry out good deeds in this world, they go to a higher, sun-filled road and enjoy the next life there. If they commit bad deeds, they travel the lower roads and suffer the consequences of their actions the next time around. Depending on their thoughts at the time of death, they are reincarnated into the appropriate place. For example, if they thought about their family, they would be born into that same family. If a person is thinking about money at the time of death, they'll be reborn as a trader or merchant. And if they are thinking negative thoughts at the point of death, they will enter a lower world and suffer in the hands of evil. A Hindu who goes to Heaven will enjoy the pleasures of the place but in the end they will realise that Heavenly pleasures are not the ultimate goal. However intense those pleasures may be, they will not last long. In the Hindu world the purpose of Heaven is to impart wisdom and detachment, and constant consumption is a false route towards the ultimate goal. [16]

A common belief to indulge

While many concepts of Heaven are common to most religions, the approach to indulgence in earthly comforts varies. Islamists, Christians, Mormons and Jehovah Witnesses all have Heavens which entertain all the indulgences of Earth. In contrast Buddhism and Hinduism encourage the person to give up these earthly desires before they can achieve enlightenment.

It's interesting how similar our commercially achievable ideal lifestyles on earth mimic so many religions' description of Heaven - food, drink, jewellery, beautiful clothing, beautiful people... there's even a coach in the Quran. In a consumer society, the idea is that you will have to work hard to eventually gain the things you desire, and a similar story has been presented by some religions in their 'sell' of the afterlife:

abide by our rules and you will be allowed into Heaven, where you'll find all the wealth you can imagine.

Links between religious worship and wealth have often been close - and have sometimes become inappropriately commercial. This has influenced the way we have related to endings and closure experiences. Many churches have benefited financially from worshippers' fear of death. Providing safe passage to a comfortable and bountiful Heaven was motivation for many religious people, who obediently donated their earnings and supported their church with gifts. Many religions have traditionally drawn income from their believers, with one of the most common methods - used across cultures for thousands of years — being the tithe. It was a regular payment, rather like a worshipping tax. Established in the Near East in ancient times, the Assyrian Dictionary from Mesopotamia reveals descriptions of tithes on garments. [Referring to a ten per cent tax levied on garments by the local ruler:] *"the palace has taken eight garments as your tithe (on 85 garments)" "...eleven garments as tithe (on 112 garments)" "...(the sun-god) Shamash demands the tithe..." "four minas of silver, the tithe of [the gods] Bel, Nabu, and Nergal..."*[17]

The tithe is still a common practice and, surprisingly, in some cases it hasn't changed in centuries. As Gordon B. Hinckley, past President of The Church of Jesus Christ of Latter-day Saints, describes it, *"our major source of revenue is the ancient law of the tithe. Our people are expected to pay 10 percent of their income to move forward the work of the Church. The remarkable and wonderful thing is that they do it. Tithing is not so much a matter of dollars as it is a matter of faith."*[18]

This investment over many years was a steady, stable way of acknowledging your commitment to the church, an investment designed to guarantee progress to the next life. But when massive shocks hit the worshipping system and chaos looms, no amount of previous good deeds by a worshipper will help. In these situations everything changes, and the steady commitment to do good, the usual methods of ending a life, and the type of afterlife we can expect change drastically. In the 14th Century the Bubonic Plague hit Europe and an entire continent was about to be challenged on what it actually meant to die.

Death brings chaos

The Plague arrived in Oct 1347, at a Sicilian port on a Genoese trading ship from the Black Sea. It was to become the largest single killer Europe would ever know. It drastically changed the way people related to life's endings, the processes they valued at the end of their lives and how they perceived the transition to Heaven. From the Plague's introduction to Europe it only took five years to kill a third of the population. Thereafter it returned every few years to terrify people all over again, before being widely eradicated in the 19th century.

Traditionally for the Catholic religion, the dominant choice of the time, death was embraced as *"a sister and friend, a welcome bridge to eternal rest"*.[19] At the end of someone's life a priest would carefully and thoughtfully administer the Sacrament of Extreme Unction to aid the departed on their journey to Heaven. The families and loved ones left behind would hold ornate funeral processions, and bury their loved ones in consecrated grounds, a process rich in thought and reflection. Families indulged themselves in the meaning of the funeral and the clergy delivered flamboyant processes to support that indulgence. Having a good funeral, after a good life, was a great help in ascending to Heaven. Closure in this sense was formalised, steadily controlled through everyone's genuine belief in the church. Where the passage to the afterlife had once been been clear, established and simple, the Plague ended this peaceful stability.

In the midst of the horror, a deep rooted fear developed in the European worshipper. Despite a lack of formal education, many were well versed in everyday religious beliefs and superstitions, believing the deadly disease was God's punishment for past sins. From this desperation grew paranoia and fear. The purging of communities became a common pastime. Heretics were burnt at the stake, and people who practised other religions were massacred. Thousands of Jews were slaughtered between 1348 and 1349 thanks to accusations that they had created the Plague by poisoning wells.[20] People were desperate for meaning and looked everywhere for blame and solutions. Amongst all this panic and ignorance, speculation about how the Plague spread was rife. One doctor at the time was convinced that, *"instantaneous death occurs when the aerial spirit escaping from the eyes of the sick man strikes the healthy person standing near and looking at the sick".*[21]

The Triumph of Death By Pieter Brueghel the Elder

The sheer quantity of funerals overwhelmed the usual systems of worship. In the town of Givry in Bourgogne in France, for example, in 1348, a Friar noted that 649 of the townsfolk had perished in September alone, while the usual annual average for the town was a mere 28 to 29. Priests were equally terrified of the disease and often refused to attend funerals, which were justified given that more than half of the priests themselves were killed by the disease. Funerals had to become far simpler, not just owing to the enormous workload, but also because of concerns about infecting the mourners.[22]

Giovanni Boccaccio, the Italian writer and poet, described the devastation at the time. *"How many valiant men, how many fair ladies, breakfast with their kinsfolk and the same night supped with their ancestors in the next world! The condition of the people was pitiable to behold. They sickened by the thousands daily, and died unattended and without help. Many died in the open street, others dying in their houses, made it known by the stench of their rotting bodies. Consecrated churchyards did not suffice for the burial of the vast multitude of bodies, which were heaped by the hundreds in vast trenches, like goods in a ship's hold and covered with a little earth."*[23]

In the horrifying years when the Plague was at its peak, between 1348 -50, it killed 20 million people. It left Europe physically and emotionally drained, and into this hole fell a death-obsessed, vulnerable population. Death, its meaning, the wishes of God and the pull of Heaven were all questioned by the average European. Art started to favour subjects of Hell, Satan and the Grim Reaper. The Catholic Church was overwhelmed, its workforce shrunk by fatalities. It went on a recruitment drive, ordaining new and inexperienced priests. On the whole worshippers weren't convinced by them, and the new priests soon developed a reputation for incompetence. The lack of leadership and erosion of reputation made way for niche groups to establish themselves and present alternative approaches to the Plague. The Flagellants, for example, took to whipping themselves to cleanse past sins, in hope that the wrath of God would end.

Perhaps surprisingly, the Catholic Church became richer at this time as so many people bequeathed their wealth to it. Religion had become a commercial business. And as any normal service provider would do when demand increased, the Church started to charge for the services they offered. Catholicism became an aggressive business practice, operating very like a religious Disney, commercialising products that were sold at a variety of price points to attract the broadest possible market. People were desperate for any product that would help them on their personal day of judgement. Relics were a popular choice, allegedly items that were closest to Jesus when he was on Earth, things like straw, feathers, relics of the saints, and even pieces of the cross in the form of chunks of wood.

Pilgrimages were encouraged and exposed the pilgrim to commercial products at their destination. Holy water, badges and certificates of proof that they'd arrived at the site were all popular. Worshippers were also expected to donate regularly to the Catholic Church through a collection at the end of each service, and through tithes at the end of each month, usually about 10% of their salary. The rich could buy high positions for their children in the church, or purchase 'Indulgences' underwritten by the Pope. These would pardon a person's sins and provide access to Heaven. A later re-launch of this particular product allowed an Indulgence to be bought for someone who had already died and was trapped in purgatory—something that happened to many people who died of the Plague and were given a quick burial without proper ceremony. The poor, in contrast, could barely pay for a christening, the recommended way to kick-start the process of being a good Catholic. For those who couldn't afford to pay money, the Catholic Church would accept animals, grain, and wide range of other products in forms of payment of tithes. If you couldn't manage that, then there were plenty of opportunities to work on church-owned farms as a means of payment. The Catholic church held absolute power over personal endings for the people of 14th Century Europe. You couldn't die without them dictating what death meant, and where you went afterwards. They owned everyone's closure experiences, enjoying a monopoly over a good death. Something had to break.

Structural off-boarding breaks down

Plague times had been the golden years of death, endings, and closure experiences. The disease had provided the biggest loss of life in history, making death an incredibly familiar issue to everyone. The Church had presented a structure and doctrine for people to follow. If they didn't follow obediently, the carrot and stick of Heaven and Hell brought them back into line. The period's obsession with death, spirituality and the afterlife had created a powerful hunger for positive alternatives. So the Renaissance came as a welcome relief, from the 14th century onwards, bringing a new approach to birth, death and what happened in between. New discoveries in science asked profound questions about our relationship to the stars, through the work of Nicolaus Copernicus (1473-1543) and his publication of *On the Revolutions of the Celestial Spheres*.[24] Instead of an omnipresent God looking down from the heavens, people now found the sun in the centre of a new Renaissance solar system.

In the 1440s Johannes Gutenberg's printing press helped spread these new ideas, providing a brand new, high tech medium for knowledge distribution. Knowledge now became more accessible to many more people than the old-style illuminated manuscripts of previous centuries. The printed word also helped establish changes in religion. Martin Luther's work *Ninety-Five Theses*, produced in 1517, challenged the Catholic church and its methods. He raised issues about the selling of Indulgences, the distribution and collection of funds, and the interpretation of the Bible. Luther believed the style of worship that the Catholic church promoted had diverged from the teachings of the true faith. He took particular issue with the Pope's distribution of Indulgences as it suggested he had an authority over purgatory, and even branded the Pope 'the Antichrist'.[25]

Luther captured these criticisms in a short essay, written in Latin, which he placed on his chapel door - a common practice at the time as a way of inviting academic discussion. Writing it in Latin helped keep the information from the masses. Initially he had no intention of inviting discussion from the wider worshipping population, but someone translated the work into a local dialect and re-printed it. One aspect of particular interest for the local people of Saxony was why they should pay for the upkeep of buildings in Rome when they had so many local needs to deal with. Luther's work struck a chord with many German worshippers. Others across Europe who had started to harbour doubts soon took up the campaign. Many Southern countries remained true to Rome's version of the faith, and they took up arms to fight the challenge of Luther's Protestant uprising from the North. During the years that followed, countless people were slaughtered in the name of God on the battlefield and across kingdoms. The ebb and flow of the Christian faith went on for decades, and ultimately established a new version of Christianity. It was a more human-centred, open faith that wouldn't be all about endings, death and Heaven. Instead, it drove its followers in an entirely different direction.

Protestantism

Thanks to the Renaissance, the Protestant world view aligned with a new interest in science and humanities. It freed worshippers to use all their *"God-given faculties, including the power of reason, to explore God's creation and, according to Genesis 2:15, make use of it in a responsible and sustainable way."*[26] A Protestant was encouraged to believe in gratitude for their redemption in Christ, resulting in a strong sense of responsibility, industry and sometimes an actual calling. Some versions of the new religion, for example Calvinism, rejected explicit luxury, encouraging craftsmen, industrialists and businessmen to invest their profits in machinery, technology and improvements to efficiency.

These new beliefs fascinated the sociologist Max Weber (1864-1920), who wrote the book '***The Protestant Ethic and the Spirit of Capitalism***'.[27] Widely considered one of the founders of sociology, Weber was the son of a Calvinist mother and politician father. He grew up watching the industrial revolution take hold in Germany, with factories and industry springing up in his own country and across Europe. He would have witnessed dramatic changes in society, and seen how his mother's faith interpreted these events. Later, as a professor, Weber studied business practices throughout the world, and thanks to his keen interest in economics, noted links between successful German businesses and the Protestant backgrounds of their founders. For him it revealed connections between psychological phenomena in the Protestant mind and functions of capitalism.[28]

Maturing this thinking in his thesis, he came to believe that Catholics had it relatively easy. They could offload their sins via a priest. Protestants, on the other hand, had to wait until their day of reckoning to be judged, since they did not believe in Catholic-style pre-cleansing. In Weber's opinion this resulted in heightened anxiety for Protestants, who had to prove themselves over an entire lifetime. They believed throwing themselves into work to show their love for God was the only way to remove the sins of Adam. This broadened the definition of the 'work of God' held by the Catholic religion way beyond the 'religious work' undertaken only by priests, nuns and popes. The new thinking was that anyone who worked hard could now do so 'in the name of God', and this drove more interest in professions and the career path mindset we have today.

The Catholic religion looks toward death, acknowledges death and prepares for it at certain times in life. The Protestant lives life every day through work. Creating new projects, re-investing money and creating a psyche of production and renewal is the basis of consumer society. It doesn't focus on endings, doesn't encourage discussion about closure, and represented a sea change in our relationship with closure. This Protestant cycle of investment in better machinery, technology and science produced efficiencies in production, which in turn drove increased financial returns that resulted in higher wages. Arguably

Protestantism was providing a framework for consumerism.

Amongst the wide ranging changes championed in the Protestant uprising was a more liberal approach to fasting. Martin Luther believed Protestants should not follow the rules of fasting set down in the Catholic Church, suggesting instead that this should be at the discretion of the individual. He criticised fasting as something that was purely external, so could never gain personal salvation. Hereby he rejected the dietary rules of the Catholic church.

John Calvin, another big influencer of the Protestant religion at the time, believed worshipers should not rely on designated fasting events, but should commit to a life *"tempered with frugality and sobriety"*, in effect living a life of perpetual fasting.

A framework of consumerism

Economies that developed in the West along the lines of Weber's observations showed a heightened interest in consumption as a mode of re-investment. In the past, increased wages might have attracted people to work fewer hours, but as the consumer society gained confidence, people wanted more disposable income, not reduced hours. In her article, **Consumerism — an Historical Perspective**, Sharon Beder notes that *"higher wages helped in this shift from the Protestant ethic of asceticism to one of consumerism that fitted with the required markets for mass production. In boom times, workers were given increased wages rather than increased leisure."* [29] As we all know, consumerism won the battle, creating a powerful upward cycle of work, investment and consumption. And our purchasing power increased dramatically across these new economies.

Alongside these religion-led changes in our approach to work came significant changes to the way we mourned the dying and the dead. The emotional approach to death evolved significantly in the 19th century. Initially we experienced a hysterical mourning, followed by a softening and silencing. These days we tend to cover death up for the benefit of society.

In previous centuries we would expect death more-or-less calmly and watch our loved ones pass away, often believing they were going to a better place. At the start of the 19th century, people started to reject the loss of a loved one, first through increased emotional outbursts, and then by wanting to keep them alive, not in a religious immortal sense, but as a physical reminder through ornate tombs. Philippe Aries, in his book, **Western attitudes towards death**, notes that *"this did not derive from the concept of immortality central to the religions of salvation such as Christianity. It arrived instead from the survivors, unwillingness to accept the departure of a loved one, people would hold onto the remains, they even went so far as to keep them visible in great bottles of alcohol."*[30]

Perhaps in reaction against hysterical mourning, a new way of thinking came later and produced even more surprising effects. It sprang from two sources and generated well-meaning ideas about easing the burden of the dying person and muting the gravity of their situation. 'It's all gonna be OK' spares people the worry that their life is at an end and not mentioning the gravity of the situation is supposed to benefit the dying. The resulting sentiment was, as Aries put it, is one of *"modernity; one must avoid no longer for the sake of the dying person, but for society's sake. For the sake of those close to the dying person-the disturbance and the overly strong and unbearable emotion caused by the ugliness of dying and by the very presence of death in the midst of a happy life, for it is henceforth given that life is always happy or should always seem to be so."* Despite little changing in the outward appearance of our rituals of death, we had already started to empty them of their dramatic impact. As Aries said, *"the procedure of hushing-up had begun."*[31]

Once upon a time death was familiar. It was a stable, an expected part of life. We knew how it worked and we built processes around it. It had a rich vocabulary that provided a construct in which to discuss every type of ending. It aided reflection and bought closure with it. Over recent centuries, we have experienced dramatic changes. The change has been far too gradual for individuals to remember, but it has proved hugely significant over many generations. We have distanced death.

First we used the mechanisms of Protestant ideals to champion work, careers, industry, technology and the reinvestment of profits. This potent mix benefitted our economies throughout the West. Providing an alternative to the bounty of Heaven, this new way of thinking meant everything you could imagine was here on earth to enjoy. All you had to do was work hard and it could be yours. Secondly, we removed our familiarity of abstinence by taking away enforced fasting as part of the religious diary. And thirdly, we've experienced the steady distancing of death. We hide it away, even from ourselves. Death has become repulsive, and we don't want society exposed to it. At the same time we have suppressed the emotions that allow us to reflect on endings. We deny closure in life, in death and in our consumer experiences. Previously, talk about death and endings would be common and comfortable. Now we shun any thought of endings in favour of narratives about life, work, and investments. Our lost vocabulary means we've crippled our ability to navigate over-consumption. Banks are now considered 'too big to fail'. Endless sharing online erodes our personal reputations. Prolonged and protracted discussions about climate change mean we've wasted the time we need to deal with it, bringing us to crisis point. We have forgotten how to talk freely about endings. We've lost the belief that we can't go on forever. And we're stuck in the status quo.

Introduction to chapters 3 & 4

"Man was born free, and he is everywhere in chains. Those who think themselves the masters of others are indeed greater slaves than they. How did this transformation come about? I do not know. How can it be made legitimate?"

The Social Contract. Jean-Jacques Rousseau. 1712-1778.

The consumer lifecycle in the workplace and home was clear and simple to the people of the pre-industrial world. They recognised the resources in what we would consider waste - scraps from the kitchen were given to the animals, the waste from the animals was spread on the land fertilising the plants that produced the food they harvested. They were in control of the system, manually managing every aspect of it, knowing when they could intervene and change its course. It was actionable, visible, and understandable to them. But over the coming centuries this relationship would be distanced. We would no longer see it so clearly. We would change our role from that of being an active cog in the lifecycle, to a passive consumer of it. From dealing with waste, to creating waste over which we had no control. This change happened in two directions.

The Distancing and the Fading

In one vital way our world view of waste, its role and its impact would change. The old world, where the waste of one item could clearly signal the birth of another, changed as a result of discoveries in science and technology that challenged our grasp of scale, revealing a new world of invisible, un-actionable waste. Our perception of waste went from the visible products of farmyard and home to invisibly small germs, radioactive matter and bacterial diseases, the infinitely big issues of deadly smog, climate change, 'too big to fail' banks, and online personal data which is impossible to fully erase. The sheer breadth and volume of waste fast became incomprehensible to the average consumer.

With this we lost the capability to define waste easily. We lost the ability to touch or control it, and in many respects lost responsibility for any part of it. The concept of waste, in whatever form, was becoming distant and fading fast. We relinquished our traditional responsibilities to local and national government, organisations and representatives, who make decisions about waste and remove it.

The Quickening and the Tethering

In addition to the loss of responsibility for disposing of waste, the basic role of the consumer has changed. Instead of being the active creator and controller of the lifestyle of products, the individual became but a cog in an infinite economic machine. The creation of more efficient sales systems, greater variety of choice and the ease of access to money have driven us to consume well beyond our needs. Such pressures have created the concept of consumption as a given, as an integral part of modern identity and obligatory commitment.

Many historic events contributed to changing this perception in us as consumers. In the next few pages we'll look at a few of these events and the impact they had on us.

Chapter 3

The distancing and the fading

14th century England had a population of about 3 million. The majority - about 80% - were in the countryside, populating villages, hamlets and farms. [1] Most worked as agricultural labourers on the land, manually sowing, harvesting and processing crops, like many generations before them.

Their relationship with waste was active and clearly visible. What was not consumed from the kitchen table was fed to the animals as scraps. In turn, their waste was captured and spread on the fields to nurture the land. The crops produced were then processed and eaten. A simple lifecycle of consumption - the waste of one thing - gave birth to

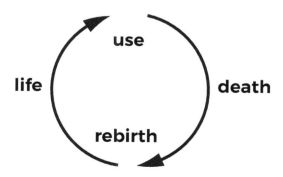

nutrients for the next. It was visible, active and understandable to the people that were consuming the produce around them.

In addition to agricultural labour, there were a variety of small industries where people scraped a living from day to day. One of the biggest of these sectors was the creation and selling of handmade goods. The craftspeople who created these products were skilled multi-taskers, taking on many roles in the manufacture of the products they were creating. They commonly worked as part of a system called 'putting-out', where supply chains were short, getting raw materials from a local supplier who would 'put them out' for multiple groups. Once a batch was completed, the supplier would pick it up and sell it though a guild in the local town.

In many cases these small production teams were part of the same family, with *"children carding the wool, women operating the spinning wheel, and men working the loom shuttles."*[2] as Michael T. Hannan, a Stanford Professor who has written widely on organisational management, describes it.

This home manufacturing, similarly to the agricultural workers, focused on a simple life-cycle of waste avoidance and recapture. They would use every resource available and if they found waste in their short production process, it was used efficiently elsewhere. They had control of, and visibility over all aspects of the production and cared deeply about the materials they worked with, a common behaviour for *"cultures based on handcraft. The stewardship of materials was a concrete element of daily life, valuing materials for the labour embodied in them and the uses to which they might be put."*[3], as described by Professor Susan Strasser.

Awareness of actionable waste was a key part of life before the industrial revolution. Consumption relied on our ability to identify waste and put it to good use elsewhere. When the materials of one process were exhausted, they were re-birthed into use elsewhere. Their consumer life was a genuine cycle, tethered to the visibility, control and management of waste. It was clear and understandable. But it was about to become obsolete.

The distancing starts

The industrial revolution punctuated the smooth valleys of Britain with its chimney stacks, railways and canals. At the heart of this new network were the emerging factories that shipped their wares throughout Britain, and in some cases throughout the world.

The factories fuelled growth in cities and swallowed up the labourers who had left the countryside. These factories crushed the small cottage industries that had previously made the nation's products. Machines now produced more items, at a higher quality, with consistent output. Which delighted traders and consumers, who'd tired of the 'putting out' system and its poor quality output.

For the craftsmen and women, though, it was a profound loss. Within a generation, their role had shifted from an empowered ownership of what they produced (and subsequently the waste generated by the industrial process) to working as hired labour on an hourly rate, doing a repetitive task. Responsibility for the waste now sat with the company they worked for, the industry that the company represented, and eventually the legislation that represented society. Although workers might not have consciously missed that role, it was the start of the distancing in responsibility for waste from the worker.

Our ancestors had visibility and control of the production in their small farmsteads and cottage industries. The following generations could only watch as the industrial machine produced bountiful machine-made waste alongside numerous products. Waste wasn't an individual problem any more for the worker.

In contrast, consumers were well versed in breaking down a product into its raw constituent parts, then re-using them to make new products, or selling the materials. It was common for women to create clothes for the rest of the family from recycled products. Men, who would often be working with their hands as tradesmen, could also be handy around the house, using their skills to mend things or creating simple pieces of furniture and toys.

The boundary between what was useful, collectable, or sellable was also visible and actionable to the consumer. Re-purposing objects, or dismantling what was not used - paper, bones, metal, wood and pretty much any component - was the norm, and these materials were captured and put into the recycling system.

The recaptured materials were also sold to the Rag and Bone man, who became the cornerstone of consumer closure experiences. He represented the boundary between use in the home and raw materials passed on for cash, exchange or even credit, brokering the boundary between consumption and manufacturing. The Rag and Bone man also handed back resources to the factories, who would start the whole process again. Rag and Bone men encouraged people to be active at the end of the consumption process. Consumers consciously ended the journey, seeing the utility in that item and breaking it apart, neutralising it into base materials again, both actionable and visible.

Action not required

However, this important off-boarding in the customer life cycle was fading. Technical advancements gave birth to more choices for manufacturers, who saw no benefit in involving the consumer in this activity. It was complicated and messy, and it didn't fit with the needs of an expanding industry. So the actionable and profitable part of the consumer lifecycle was taken away from the consumer with the loss of the Rag and Bone man.

Susan Strasser chronicles the demise of their role in this passage from her book, *Waste and Want*. "*New paper making technology is substituted wood pulp for rags. Mechanisation and, later, Prohibition destroyed that used bottle business. Swift and Armour (American meat packing companies) produced and sold enough bones to put an end to collection from scavengers. The giant modern meatpackers marketed by products to fertiliser companies and other firms that required massive amounts of skin, hair, and* bones; they also produced their own fertiliser, glue and other products that are used animal waste."[4]

Parts of the customer lifecycle were being pushed further from the visibility of the consumer. And the consumer role was gradually becoming biased towards purchasing, towards on-boarding. The balance of the customer lifecycle tipped even further toward consumption, and even further from responsibility, and off-boarding was soon taken out of the cycle of consumption altogether.

Boundaries

As the changes in actionable waste removed our ability to handle, break apart and control what is useful and what is not, we distanced ourselves from the presence of waste. Where we once tolerated our own mess, we have now framed it, defined it and distanced it, disconnecting it from our consumption. Susan Strasser explains the thinking process we go through. "*Sorting and classification have a special dimension this goes here that goes there. No trash belongs in the house: trash goes outside. Marginal categories get sorted in marginal places (attics, basements, and outbuildings), eventually to be used, sold or given away.*"[5]

This changing definition of waste reflects a history of boundaries between the consumer and the consequences of their consumption - storing it, using it, or disposing of it. We establish rubbish at the boundaries of our society. That boundary transitions, out of the house, beyond the yard, to the street, to the city wall, to the sea, to the next state.

The street was a popular place to dump our waste for many centuries. Before the 19th century, in what are now advanced developed world cities, it would be common to see scavenging animals thriving on the scraps. Dogs, pigs and goats were common aids in waste management. Even vultures were seen as a vital part of the cycle of waste in West Virginia, which in 1834 prohibited anyone hunting them for that reason. [6]

Another common boundary was that between land and water. Very like a climate change discussion today, people never really thought they could damage something as big as an ocean. So it became a favourite place for us to dump rubbish. Streams, ponds, and rivers were also easy prey for the remnants of our

Rubbish barges dumping in the sea. Scribner's Magazine 34 (October 1903). The Johns Hopkins University Press, 2000

consumption. As items faded beneath the water they could easily be forgotten.

Hiding our waste in the waterways around us proved a short term solution. And when it returned to haunt us it was no longer the physical, visible waste we were familiar with.

Invisible vulnerabilities

Human knowledge expanded with the technological and scientific advances in the 19th and 20th centuries. We needed to use that knowledge to redefine the scale of the boundaries that defined our waste. The physical ability to see was no longer enough.

People had long made assumptions about how diseases spread. Religions told us that sinners deserved disease. Some believed that 'foul air' spread 'germs' - and let's face it, there was plenty of foul air around in earlier centuries. The physician and researcher John Snow had no time for these stories, believing instead that something other than sin was to blame. In 1854, an outbreak of cholera in London's Soho provided Snow with an opportunity to investigate the source of the disease and prove his theories.[7] He tracked cases on a map of the area, showing the daily routines of the people involved and their common water source - a pump on Broadwick Street. The pump proved to be the source of the outbreak, having been polluted by an overflowing cesspit. On his advice, the authorities removed the pump handle and local instances of the disease dropped dramatically, proving to many that Snow's diagnosis was correct.

The concept of germs - and with it a new definition of invisible waste - took some time to sink in. The public tended to prefer the simpler explanation of God's will, or bad air. Even amongst Snow's contemporaries, germ theory wasn't widely accepted until the 1860s.[8] However it did eventually change our perception of dirt or waste to something more complex, something that wasn't easy to see or touch, which was difficult for people at the time to understand.

Mary Douglas, the prominent anthropologist and author of **Purity and Danger'**, points out how this changed our modern perception of dirt or waste, as it became *"...dominated by the knowledge of pathogenic organisms. The bacterial transmission of disease was a great 19th century discovery. It produced the most radical revolution in the history of medicine. So much has it transformed our lives that it is difficult to think of that except in the context of pathogenicity."*[9]

This new definition of waste gradually widened. Douglas argued that this redefined the perception through symbolism, concluding that removing the science left us with two conditions for a new, broader idea of waste, a *"set of ordered relations and a contravention of that order."*[10]In other words, 'matter out of place'; things like cake on the mantelpiece, dirty cups in the bathroom, soup stains down your tie. The issue was one of organising objects in a context that we felt comfortable with or, as Douglas describes it, an *"object or idea likely to confuse or contradict cherished classifications."*[11]

...and the kitchen sink
The introduction of the the waste disposal unit allowed many

consumers to redefine the boundary of waste in the kitchen. It also gave kitchen manufactures an opportunity to sell the distancing of waste as part of a new American domestic dream.

Previously, the stewardship of waste in the kitchen had presented an opportunity for separating and storing useful items that could later be made into new meals. For some, waste disposal introduced a compelling convenience that few had ever had before. For others it was introduced as an improvement to domestic hygiene. The man of the house welcomed this new technology, since taking out the trash was one of the few domestic roles they were expected to take responsibility. District councils even saw the benefits. The city of Jasper in Indiana, for example, halted the collection of food waste, believing it was the source of an outbreak of cholera in the city. Instead they chose to introduce waste disposal throughout the district, and put all waste into the sewage system, ending the unpleasant exercise of separating different types of waste. The boundary for waste was now the plug hole and the kitchen sink, with the waste being sent into the sewer, invisible and distant.

With this, the customer lifecycle had changed, reducing the previously-mucky off-boarding of food waste to a mere moment spent shoving it down the plug hole. This enabled the consumer to enjoy the delightful experience of their new kitchen, serving meals for the family, living the modern dream of domesticity - a cherished story at the time in America.

Away from the sink and food, other kitchen items were changing in response to new packaging techniques and materials technology. Consumers were enjoying new-found convenience and they were looking for it everywhere. The introduction of throwaway items such as paper towels, cups, straws, and the common use of commercial toilet paper sealed the trend.[12] Companies like Kleenex emphasised to consumers how good it was to have convenience in waste. The end of the consumer life-cycle was getting quicker, easier and ever-less meaningful.

The invisible fears of waste

As scientific knowledge about germs grew, knowledge in the general population increased slowly. Germs were added to a fast-growing list of invisible fears. The germs had the power to do all sorts of bad things, something that was soon exploited by selling a whole new genre of products based on killing these invisible germs. Consumers sucked it up fearfully, desperate to halt germs in their tracks before they damaged the family's health, a picture-perfect paranoid product opportunity.

These new products provided very little visible evidence for the consumer to gauge success when cleaning. Wiping a surface with a cloth dampened with water, or with the best selling cleaning brand, would have a similar outcome to the eye. People had to trust the manufacturer and believe expert opinion, something that advertisers started to employ skilfully. In no time they were selling the solution to fear via new products, and we found ourselves in a 'post-visual' world of waste.

The haunting waste of radiation

If germs worried consumers, the waste left behind by nuclear radiation terrified us. The earliest of works by Wilhelm Röntgen set a precedent regarding nuclear's haunting, magical

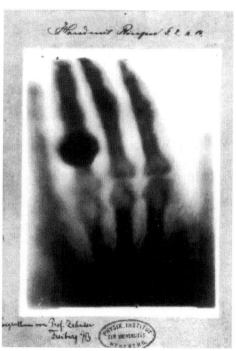

First medical X ray. Wilhelm Röntgen.
via Wikimedia Commons

power when, after experimenting with cathode rays, he observed some photographic plates he'd energised giving off a faint glow.[13] For the following two weeks he observed and experimented with this weird glow, culminating in the first X-Ray photo of his wife Anna Bertha's hand. In a shocked reaction to seeing the skeleton of her own hand, she exclaimed, "*I have seen my death*"

Germs were fairly easy to understand compared with nuclear radiation. Awareness of germs had been increasing thanks to public health messages and adverts. Nuclear waste was a whole different issue. Not only was it invisible, it was complex, a further step away outside the understanding of the common consumer, who saw it as terrifying thanks to associations with the atom bomb.

Only government experts could deal with this waste. Regular people were powerless to touch it, see it, understand it, or do anything about it, and we were castrated from action. All the consumer could do was sit back and worry. It was another nail in the coffin of actionable off-boarding, and reinforced the new message that regular people can't do anything about endings - you need to be a specialist, a government or have the right equipment.

Biggest of boundaries

In the 1960s consumers had some pretty big things to get their heads around. And they didn't come much bigger than the moon landing. There is a certain irony in that one of the most polluting individual acts of humankind — reaching the moon — gave rise to something that changed our perception of waste upon the Earth.

Apollo 8 launched on the 21st of December 1968.[14] It was the first manned flight to leave the earth and orbit the moon. The Saturn V engine of the rocket burnt an astonishing 13 metric tonnes of propellant per second for those two and half minutes, taking the rocket to the heady height of 42 miles and a speed of 6,164 miles per hour, dumping 2,100,000 kg of rocket fuel into the atmosphere.[15]

Its human cargo was three astronauts, Frank Borman, James Lovell and William Anders, who were about to become the first people to exit a low earth orbit and see our planet as a whole for the first time. The significance of this only became apparent from an interview of the crew on the evening of

Earthrise. Apollo 8. Nasa. Dec 24th 1968

Christmas Eve a few days after the launch. While doing a live broadcast, they showed an image taken by William Anders of the beautiful blue earth, now considered one of the most historic photographs of the 20th century - Earthrise. [16]

James Lovell reflected on this poignant moment, saying, *"The vast loneliness is awe-inspiring and it makes you realise just what you have back there on Earth."* [17] and Earthrise was hailed as *"the most influential environmental photograph ever taken"* by the nature photographer Galen Rowell. [18]

The photo ultimately delivered a kick start to the environmental movement, highlighting what a fragile and humble planet we live on; its clouds, seas and land revealed like a precious jewel glowing in the lonely darkness of space. This one photo alone redefined the boundary of our home planet, forcing humanity to step back and recognise our role more clearly. What was usually invisible was now visible from this new angle, and it was inspiring to everyone, broadcasting a universal message about human achievement and community that everyone could understand. Earthrise inspired self reflection and, for a while, we all started thinking a little bit about who our neighbours were and what impact we might make on what suddenly looked like a delicate earth.

Many ecological groups were formed in the aftermath, including Earth Day, Friends of the Earth and The Whole Earth Catalogue, all of which emerged soon after the moon landings. Archibald MacLeish, the modern American poet,[19] reflected upon earth's new-found human community in *The New York Times*, "*brothers who know now they are truly brothers... riders on the Earth together*".[20]

For a brief moment our 1960s selves were inspired to consume in a more considerate manner for the benefit of all of us and the planet itself. The revelation that our earth is fragile and vulnerable caused some of us to reconsider whether the consumer boom that took over along with the industrial revolution caused long term damage which should be avoided. This new hope and reflection on the nature of consumption was short lived. The direction of these environmental campaigns was multi-threaded and on occasions disorientating for the consumer. But within just one generation, a global threat had emerged that consisted of a single example with a global boundary, clearly revealing the consequences of long term, thoughtless consumption.

Carbon, the party pooper

Scientists had talked about the potential for climate change in the past. Some discussion of greenhouse gases took place as early as the 19th century,[21] but it wasn't until the 1990s that computer modelling and increased observational work brought consensus and climate change was born, the biggest threat to human survival identified so far in history. To deal with it, the Conference of Parties (COP) was established. This was an event attended by countries in the UN which assessed progress in dealing with climate change.

During the first meeting in Berlin in 1995, concerns were raised about the abilities of countries to meet commitments made around climate change. At pretty much every event since then commitments have been avoided, or made and broken, or redefined later thanks to the skewed logic of party politics, and by countries promising short term economic gains to voters. This has happened despite the world's population looking on in bewilderment. And now we're at COP21.

The COP21 summit was hosted in Paris. The world's leaders attended proudly with good intentions. A lot of pressure had been put on delegates, and agreements were eventually rattled out to a great fanfare. We have been here before. And we can forgive the world watching pessimistically as the summers get hotter, we suffer more autumn floods, and the winters see either freezing extremes or highly unusual warmth. The Climate Tracker website, which captures the actions and inactions of governments, puts it like this: *"There remains a substantial gap between what governments have promised to do and the total level of actions they have undertaken to date. Furthermore, both the current policy and pledge trajectories lie well above emissions pathways consistent with a 1.5°C or 2°C world."*[22]

It's no surprise that these agreements have been broken in the past. We no longer have a common vocabulary around endings, and we've been pushed way beyond personal, meaningful attachment to the decisions that need to be made. We relinquish responsibility to governments and organisations to control our waste, and with that have removed the direct connection to us. Now waste and responsibility is proxied to others, who often have a mandate for economy, not ecology.

For consumers this is a distant issue. They see an endless round of climate change discussions in the media. They hear the evidence from scientists, they recycle, they might even attend marches to encourage their leaders to 'do something!', but these are far removed from their daily routines of consumption. What damage we might do as consumers by driving, flying, turning up the heating, eating steak sandwiches and so on, does not connect to the wider impact of those actions. Circles of concern and influence are also distant. Likewise, the distance between how we begin a consumer experience and how that might ultimately end is usually invisible. We broke the link long ago. We removed the off switch. And now we don't know how to stop.

Too big to fail... see, understand, or deal with.

Beyond the physical world of carbon, products and the natural environment is the intangible world of human-created services. Financial services being one of the biggest and most influential. 'Too big to fail', was the phrase that justified the trillions of dollars/pounds/euros/kitchen sinks, that we chucked at propping up the global banking system after its collapse in 2008. The phrase alone characterised a psyche that we've indulged ourselves in for decades, that of avoiding the end the lifecycle. It reinforces the attitude of taking on debt instead of paying debt off. It goes against all the fundamentals of economics.

Adam Smith, the godfather of economics, championed the consequences of failure in economics in a man-made business evolution of survival, where the fittest company wins and the weaker company fails in a market free from outside intervention. Some would say that the real world of international finance is a little more complicated than Smith's view. It certainly seemed that way in 2008, when people realised we didn't understand the system we'd created, and that we'd actually created a monster. A system that, apart from being 'too big to fail', was also too complicated to see, too complex to grasp, too good to be true, and too focused on the short term. Based on greed, it was a biased customer experience with lots of on-boarding and no way to off-board.

Paul Volcker, the former head of the Financial Reserve, noted in a speech in 2008, just after the crash, that on re-reading Adam Smith, the passage about Scottish Banks and their size was pertinent to the situation we're all in.

Smith said some banks in Scotland had grown far too big. A healthier situation would see many smaller banks created, to soak up the failure of these big banks, as Smith said, by *"dividing the whole circulation into a greater number of parts, the failure of any one company, an accident which, in the course of things, must sometimes happen, becomes of less consequence to the public. This free competition, too, obliges all bankers to be more liberal in their dealings with their customers, lest their rivals should carry them away. In general, if any branch of trade, or any division of labour, be advantageous to the public, the freer and more general the competition, it will always be the more so."*[23]

Despite Smith's recommendations being 230 years old, and so widely read by the financial establishment, we still seem to fail at implementing their lessons. Volcker noted that many of the smaller banks were absorbed by the bigger ones before 2008, so there wasn't a Smith-style safety net available. As a result we all fell into the worst financial disaster since the 1920s.

The Credit Crunch had many sources apart from the fundamental failures Smith noted. In a rapidly rising market, with historically low interest rates, ordinary people started borrowing money. Excited about getting on the property ladder, they were sold the dream by overwhelmingly confident advertising about loans, on-boarding buyers to the mortgage they dreamed of in no time, without the usual checks and balances. And millions of people borrowed way beyond their means.

Lenders had become pretty relaxed with who they lent to - which was almost anyone. 'Sub-Prime' mortgages seemed to provide a solution for everyone to borrow. Banks then bundled these sub-prime loans together and their financial engineers turned them into products to resell. One of these, a 'collateralised debt obligation', aka CDO, became the ground zero of the problem, assembling bad and good loans into one package. The trouble was, the sub-prime loans started to dominate CDO products. And nobody really knew what was in the CDOs. Was it 50% bad sub-prime mortgages, or more like 95%?

Again, in the intangible world of services, we see the waste in the system being hidden and made un-actionable. A dreamy, persuasive story told at the beginning – the on-boarding – had no feasible middle or end – off-boarding - for millions of regular borrowers. The bias in the balance of the customer experience was extreme. The only part of the customer life-cycle that consumers themselves could actually action was to sign-up to the loan. Beyond that, it failed.

There are also similar boundaries of waste to the physical product world evident here. The simple historic relationship of the banker lending money to the borrower has been broken into many other participants: the financial reserve, the bank, the financial engineers, the broker, the agent. Each of them interspersed a new delusional boundary in the long chain beyond the personal responsibility of the consumer and the ending. The distance between them was enormous, too far to provide empathy to the borrower and too far to provide clarity in the product.

The Economist wisely highlights all our roles in a financial crash. *"And as so often in the history of financial crashes, humble consumers also joined in the collective delusion that lasting prosperity could be built on ever-bigger piles of debt."*[24]

The digital lingering legacy

The human-made world of digital exists behind the screens of our computers, rendering the services and products we use in a landscape of pixels. In its 40 years of evolution, many think it is our greatest ever invention. And on lots of counts it is. The digital world is often championed as a democratic tool and universal leveller. Hillary Clinton said it *"allows individuals to get online, come together, and hopefully cooperate. Once you're on the internet, you don't need to be a tycoon or a rock star to have a huge impact on society."*[25] This has particular poignancy in our so-called post-truth world, when a property tycoon beat her to the White House.

In this digital world, our creative output has been enormous. Cisco believes that *"Annual global IP traffic will pass the zettabyte ([ZB]; 1000 exabytes [EB]) threshold by the end of 2016, and will reach 2.3 ZB per year by 2020."[26]* A Zettabyte is 'a unit of information equal to one sextillion, or 2 followed by 70 zeros, bytes.'[27]

These creative acts are captured through short, flippant messages on Twitter, rambling essays on personal blogs, funny cat pictures, dull lunch pictures, important videos of far away revolutions, casual erotic pictures sent to partners, and meaningless rants sent to everyone - important information for all. The internet captures, holds, shares and stores all of this content.

What the internet doesn't do well is un-share. The waste we create on the internet, the things that we thought were important at one moment, quickly seem unimportant, wrong, embarrassing or out of date the next. In different situations this would be categorised as waste, but online it still retains an accessibility that we wouldn't experience in the world of waste, whether products or services. In fact it is almost impossible for a wide variety of reasons to completely remove anything from the internet.

Great interface design, interfaces that encourage input, action, collaboration and involvement from the user and their 'friends', overlooks the need to detach, remove and take personal responsibility. Instead it often leaves the responsibility of consumption to the governing body of the service provider, who also can't do much for you if you have unwisely shared the wrong thing.

The waste product of consumption now has no actionable end at all. It is left lingering on the internet, waiting to undermine the individual's reputation. Now we have to create laws, like the 'Right To Be Forgotten', to allow for removal of the content we've 'over-shared'. It isn't that we can't create user-focused tools that let individuals control their media. We've been so far removed for dealing with waste/endings/closure as individual consumers that we've forgotten how to even contemplate that it might be a requirement.

The consumer journey from the 14th century has been a

Ends.

wonderfully progressive experience. We have benefited constantly from the creation of abundance in products, services and more recently digital. But as we indulged in this bounty, we gradually distanced ourselves from the importance of responsibility and reflection at the end of the customer lifecycle. We chipped away at it slowly and efficiently, generation after generation. Now the off-boarding of a consumer experience seems totally alien to any other aspect of our consumption. Off-boarding has become far too complex, too inconvenient, too big, too fast, too hard to deal with. And too distant.

Chapter 4

The quickening and the tethering

Alongside the distancing from, and fading effect of waste on our consumer behaviour, a quickening of consumption spread though our society from the 15th century onwards. This started in the cities and removed many barriers to purchasing. In the majority of cases it achieved great leaps forward in convenience for consumers and an enormous bounty for businesses, industry and society.

To encourage this acceleration, marketers increasingly tethered consumers' identity to their purchases. Let's explore some of these changes, their consequences for the consumer and on the way we consumed.

Increased wages

The benefits of wealth blessed few people in 14th century Britain. Much of the available wealth had been inherited from previous generations, much of taken by force.

Eirlys Roberts, writer of **Consumers**, describes the background to wealth at the time.

"In England, some of the great landowners inherited their land - from which they derived wealth and power and social status - from ancestors who

had been given it for their services by grateful monarchs. As many, if not more, inherited it from ancestors who had taken it by force - in other words, stolen it - from whoever owned it at the time."

These lucky few inheritors were at the top of a long chain of privileged consumption. Their desires were driving whatever consumer experiences existed at the time, usually for unique one-off products, handmade by craftspeople, everything from cabinets to clocks, paintings to porcelain, dresses to drapes. The production of these goods generated a trickle of money from the rich to the wider poor. The wider population, whose poverty-stricken life style was in complete contrast to that of their masters.

Slowly, the economic situation started to improve. Wages doubled between 1550 and 1650, stimulated by international trade and the high rate of inflation that occurred during this period across Western Europe. In fact, according to a paper by Nuffield College Oxford about the building craftsmen of Europe, wages had been rising for some time across the whole of Europe.[1] While every European city was feeling the benefit, London started to edge ahead, and over this period grew larger by far than other European cities, leaving London as the leading wage payer in Europe. London's success empowered an emerging consumer base which laid fertile economic ground, alongside the growing expertise in science and technology that led to the Industrial Revolution to taking off in the UK. London's population would become the new consumers, and its capital became a financial hub.

Curating the consumer

Consumer behaviour needed to change if it was to fulfil the needs of an emerging industry. In past centuries, as a consumer, no week would be complete without a lengthy journey to the market. Or, if you were lucky enough to live in a large town, visiting numerous trades people to get your weekly shop. The experience would often entail personal exchanges with proprietors who might know us by name, or at least were familiar with us. This relationship acted as a bond in expectation of a certain quality, something the proprietor upheld. Your identity as a regular consumer was important, connected only through the memory of the proprietor of the shop, whose probable illiteracy would have prevented any other kind of identity being captured.

The routine of shopping cost time and effort. Manual production processes meant it was was limited in consistency and quality. Supplies were irregular and restrictive because production cycles were short. The idea of consumers being able to choose from different suppliers seemed absurd at the time, and this traditional way of buying restricted the pace of consumption considerably. Desires were still a long way from today's obsessions, since most consumers bought goods based on real 'needs'. They didn't make the complicated value judgements we make today based on our 'wants'.

As the Industrial Revolution rolled onwards and the factories and machines started to flood the market with goods of consistent quality, choice and a steady supply chain, shopping started to change. Although change was slow for the majority of consumers, who had little consumer influence at the time, a fast-growing, wealthier, middle class of consumer started to dictate a faster pace.

These consumers demanded new approaches to buying things. They wanted bigger, brighter stores, where they could shop for all types of luxury items in one place. The first of these new 'department stores' is widely believed to be Harding, Howell & Co's Grand Fashionable Magazine, opened to great fanfare in 1796 on Pall Mall in London.[2] It was divided into four departments offering items to fulfil 'the needs of fashionable women', and removed the need to visit multiple vendors. New shops like this speeded up the shopping process and often removed the burden on the consumer who wanted to create the correct fashionable look themselves. This was now done via the creative curation of those who had selected which items to put on sale.

Aspirational products

To maximise factory capacity and keep up with consumer demand, the wide-ranging desires of consumers needed to be influenced and directed. Traditionally the marketing of products was about basic availability, something people had never experienced in this new environment. There had been little competition for many products, so not much need to talk about detailed qualities or benefits.

As markets grew and competitors emerged, marketing had to move to a sales-led approach, communicating with customers more directly to persuade them of a product's superiority. An early pioneer of this increasing sophistication was a Liverpool-based tobacco brand. In 1848, Cope Bros & Co, launched by Thomas and George Cope, was a large company making a variety of products from imported tobacco.[3] These were targeted at different markets according to class. Through advertising and packaging, the upper classes were promised the product was *"delicately fragrant"*. A more health conscious middle class smoker was told *"smoke not only checks disease but preserves the lungs"*. And a *"rugged heavy taste"* was promoted to sailors, soldiers and working men.

Framing our consumption in these terms reinforced consumers' perceptions of themselves, broadly linking identity and consumption. Taking things further let marketers bring aspiration into the equation, encouraging people to define themselves through their consumption. The work of Thomas J. Barratt, (1841-1914) is the textbook example of the time. His famous work with Pears Soap highlighted the way people were consuming as a mode of defining themselves.[4] When they bought a particular product, they fulfilled their dreams. Pears Soap was sold as an aspirational product which hinted at belonging to high society, harnessing images of middle class children on the packaging and in the advertising. The brand famously used the painting Bubbles, by the artist John Everett Millais, in one of the ads. He later complained about it, but to no avail since Barratt had bought the copyright.

The approach in these ads proposed levels of comfort and cleanliness that were far from the reality experienced by most of its 19th century customers. Many welcomed a moment of emotional escapism when using a product, allowing them to dream about and believe in a gentler, easier life than their own. The subconscious affect of such advertising made consumers forget about their everyday cares and encourage them to imagine themselves as another, more upbeat person.

Behaving aspirationally

This type of self definition through products was observed by the American sociologist Thorstein Veblen at the end of the 19th century in his book *Theory of the Leisure Class* (published in 1957).[5] He'd observed the emergence of the nouveaux riches who'd profited from the industrial revolution rather than from inheritance. He believed that they demonstrated the possession of this newly-acquired wealth so as to guarantee a place in society by impressing others with their social power and prestige. He based this theory on their leisure behaviours and the quality of goods they purchased.

Veblen believed that people earned social status by displaying patterns of consumption, rather than their true financial power. This, he proposed, influenced people in other social classes to copy this behaviour, and consequently drove everyone to buy aspirational products. As this passage from his book points out, he believed the consumer had lost all sense of responsibility and reflection about the consequences of their consumption, because they were aiming to maximise their perceived identity through what they consumed. As he said, *"The quasi-peaceable gentleman of leisure... not only consumes... beyond the minimum required for subsistence and physical efficiency, but his consumption also undergoes a specialization as regards the quality of the goods consumed. He consumes freely and of the best, in food, drink, narcotics, shelter, services, ornaments, apparel, weapons and accoutrements, amusements, amulets, and idols or divinities. ...Since the consumption of these more excellent goods is an evidence of wealth, it becomes honorific; and conversely, the failure to consume in due quantity and quality becomes a mark of inferiority and demerit.*"[6]

This approach intensified as it crossed the Atlantic. No longer restricted to the behaviour of the upper classes or the nouveaux riches in Europe, the American economic boom influenced a far wider range of consumers. Veblen observed consumption limited to a display of wealth and abundance by the upper classes, which had some influence on the wider classes, but this never achieved the disposability and turn-over that took root in the mindset of the American consumer.

The author Julia McNair Wright described the European approach to the frugal life, imposed through war and centuries of budgeting, *"the shops expect to sell in littles: a penny's worth of this and two pence worth of that. Exactly what is needed for use is bought, and there is nothing to be wasted"*[7]. She went on to contrast the phenomenon with her own American experience of being brought up on abundance, thanks to the huge land mass that they inhabited. *"Lavish abundance of common things surrounded our ancestors and they used it lavishly: we inherited this prodigal habit"*[8]. This dramatic contrast across the Atlantic was reaffirmed in a **Good Housekeeping** article of 1885, repeating a common French saying that *"a French family could live with elegance on what an American housewife throws away"*[9]

By the 1920s, against a background of financial booms typical in those early years, a new and truly American approach to consumption was emerging. Christine Frederick, a pioneering home economist with a background in advertising (who was also credited with standardising the height of worktops and kitchen counters), published the book **Selling Mrs. Consumer**, in 1929.[10] This challenged the widely held assumption of the male as the lone consumer, establishing the importance of women as informed consumers and encouraging them to buy more goods.

Sign, Birmingham, Alabama. Rothstein, Arthur, 1915-1985, photographer

She saw a clean break with the consumer habits of the old world Europeans, who thought it wise to buy once, buying items of good quality that last and last. She believed this slowed the rate of progress, and she encouraged the American consumer to engage with the 'Progressive Obsolescence' she observed in her research into the American home. In her eyes the characteristics that most defined the American approach were first a suggestible state of mind, eager and willing to take hold of anything new. Second came an innate readiness to scrap or lay aside an article before its natural, useful life was completed in order to make way for a new and better thing. And third came a willingness to spend money, a very large share of one's income, even if savings had to be cut back or even abandoned, in order to have new things and new experiences.[11] By doing so she encouraged consumers to reject the quality/longevity approach.

In Frederick's disapproving words, "*In Europe people buy shoes, clothes, motor cars, etc., to last just as long possible. That is their idea of buying wisely. You buy once and of very substantial, everlasting materials and you never buy again if you can help it. It is not uncommon for English women of certain circles* to wear on all formal occasions, the same evening gown for five or ten years."[12] Instead she championed the purchase of goods on the basis of wanting the latest, newest, brightest and shiniest thing. "We have subscribed whole-heartedly to the consumer idea that goods should not be consumed up to the last ounce of their usability; but that in an industrial era Mrs. Consumer is happiest and best served if she consumes goods at the same approximate rate of change and improvement that science and art and machinery can make possible."[13]

Few people can claim to be a bigger champion of the American consumer than Christine Frederick. She was a big supporter of gender equality, and an enormous promoter of capitalism. Her definition of the American consumer, and the rules, encouragement and guidance she provided, established the character of the time.

John Kenneth Galbraith re-framed this abundance in his book *The Affluent Society*. Pointing out the Western economic model was based on the conventional wisdom of economists such as Adam Smith, Ricardo, Malthus, Marx and Marshall, whose views were based on a post-industrial revolution world, that assumed a necessary inequality between the few rich and the many poor. This was because unpredictable trade depressions could happen at any point. This would inevitably drive the majority of people below the level of subsistence - though this was based on the assumption that not enough was being produced to supply everyone's needs. Galbraith believed that much of this previously felt inequality and insecurity had disappeared by 1950 and claimed America was producing too much, not too little. Employing salespeople and advertisers to persuade the population to buy goods that are not needed and overlook the public goods (schools, hospitals) that are needed.

"In an atmosphere of private opulence and public squalor, the private goods have full sway" stated Galbraith.

Product euthanasia

A hungry consumer with an appetite for new products had been established in America. There was a healthy rejection of old-school traditionalism. Now the dots had to be joined, and industry needed to step up to supply the goods. But there was a spanner in the works of this plan, namely the legacy of goods that were built to last, something consumers in previous decades had valued and demanded. The trouble with products lasting for years is that they rarely need to be replaced. Although this is what the consumers straight-thinking head wanted, it was not what was going to inspire their emotions or keep the new consumer boom alive and kicking. The traditional European approach of good quality was outdated, so US manufacturers developed a new model where shorter product life spans were the norm.

Martin Mayer, a writer about advertising, read an article about 'product death dates'. Inspired, he attempted to define 'planned obsolescence' in an article for *Dun's, Review* a general business magazine of the time.[14] He stated that three types of 'planned obsolescence' can be observed.

The first was 'functional obsolescence', for example frost free freezers and increasingly wide TVs. The second was 'planned failure of materials', which he pointed out is difficult to research as it quickly becomes such an emotional subject. The light bulb is often used as an example. When several companies, including General Electric, set up the Phoebus Cartel in 1921 to agree limits on the lifespan of light bulbs, they standardised a bulb's useful life at 1000 hours.[15] They even fined companies who produced longer lasting products. And thirdly, Mayer proposed 'style obsolescence', something very few sectors achieved on the scale of the US motor industry.

The appetite of US consumers until then had been for reliable, practical products. The Ford model-T, launched in October 1908, represented many of these needs. It was big enough to get a family in, but small and efficient enough for an individual to use daily. It was intentionally simple so that owners felt empowered to maintain it, and mend it when it broke. It was a great success, and remains one of the biggest selling cars of all time. Ford's Model-T was presented as a long term, practical investment echoing a type of consumption more in tune with, say, investing in a good tractor. It certainly characterised a more European approach. Although a big seller, it wasn't the star of consumption that embodied the US over the coming decades. That role was left for others.

While Ford had great success in minimising costs for the Model-T, it lacked vision in introducing new products. Their slowness to market was a reflection of limited flexibility in their factories. Take the painfully slow changeover from older factory tooling to re-tooling for the new Model-A car, something that saw them shutting down for more than six months.[16] The General Motors strategy was a little different. It gave birth to a new normal for the consumer, characterising the style obsolescence outlined by Martin Mayer.[17] GM introduced new features and styles more frequently than the lifespan of the product, inducing a disturbing consumer feeling that their current purchase was becoming outdated. General Motors designed and planned incremental changes at the heart of their products. So when updates to engines, bodywork and styles were implemented, they caused the minimum impact on production. For the 1929 Chevrolet engine upgrade from 4 to 6 cylinders, General Motors shut their plant for just three weeks.[18]

On the back of this inbuilt policy for constant changes, General Motors drove (excuse the pun), a consumer boom of desire. Because the manufacturer was geared up for change, they were able to offer body and style changes to consumers every year, and people lapped it up. At the same time consumers were subject to an onslaught of advertising encouraging competitive consumption between the brand owners - the Buick driver was doing better than the Chevrolet, but they both envied the Cadillac driver.

Daniel J. Boorstin describes the phenomenon in his Pulitzer Prize winning book *The Americans* as coming *"closer than any earlier American institution to creating a visible and universal scheme of class distinction in the democratic United States of America"*[19]

This cultural appetite for change has successfully embedded an ethos in consumers to expect continuous new offerings. Our comfort in the long term stability of a product changed to boredom. The values of quality and longevity are now often packaged as nostalgia or heritage. And the consumer lifecycle has tipped significantly towards on-boarding, more starting experiences, more new products. In its wake the consumer has been distanced from responsibility, longevity and closure experiences. On-boarding boomed at the same time as off-boarding was left in the shadows.

Cards for cash

A similar step change was required with the fundamentals of purchasing. Up until the 1900s the way money was lent was well established, with almost all consumer transactions done in cash. After the 2nd World War annual consumer spending grew rapidly and by 1949 the availability for credit was having a greater influence on consumer habits. Forty nine percent of new cars, 54% of used cars, 54% of refrigerators, and 46% of TVs were sold on credit. [20]

Accompanying this were marketing-led innovations like loyalty schemes, with businesses introducing store cards and oil companies creating courtesy cards to use at gas stations. None of these early versions were explicitly linked to a credit card, but it wasn't long before they were brushing shoulders.

The first foray into credit cards was called Charg-it, introduced by John Biggins of Biggins Bank in Brooklyn, New York in 1946.[21] Their bank card was used in stores by the bank's clients. The bill was sent to the bank, who would pay for the purchases on the customer's behalf and later settle up with them. But the Diners Club card opened the floodgates, as this was considered to be the first practical credit card.

An innovation of necessity, the idea occurred to Frank McNamara while dining with a friend at Major's Cabin Grill in New York in 1949. On getting the bill he realised he had forgotten his wallet. Reeling from the embarrassment, he decided someone needed to create an alternative to cash, to pay for meals. In February 1950 he returned to the Grill having organised a line of credit, using his new card to pay for the meal, an event known in the Credit Card industry as 'The First Supper'.[22] In the following year he distributed 200 cards and agreed accounts with 27 restaurants in New York. The number of Diners Club members grew successfully in the following consumer boom years, and by 1951 there were 20,000 members.

One of the first Dinner Club cards. 1952. history1900s.about.com

American Express, who had been in the postal and money orders business since 1850, bought their own version to the market in 1958, and introduced the first plastic card the following year.

The credit card model works in a 'closed loop' system where the issuer has full control of the transactions and settles directly with the merchant and consumer. In the same year that AMEX introduced the plastic card, they also introduced the option of a 'revolving balance', meaning the card holder no longer had to pay off the bills at the end of each cycle (usually a month), giving them additional freedom to spend.

For earlier consumers, consumption beyond the essentials usually required a level of fore-thought, either by saving for a particular item or preparing to justify a purchase to the bank manager for a loan. With credit cards consumers could borrow on a whim, without acknowledging the reasons behind the purchase. This had a significant impact on the economy over the coming decades. The 'revolving balance' feature reduced the need to reflect in other ways about a purchase. It reduced the need to end the debt at the end of the month. It removed the off-boarding of the debt experience. It encouraged more on-boarding, more debt, more denial. Paying by credit card removed the need to be a responsible debtor.

Accelerating beyond the physical

Physical limitations around consumption, for example availability of goods and their location, are some of the biggest barriers to overcome. Technology has often broken these down with innovations, like printing, a postal service, or the railroads, all of which have allowed businesses to change their relationship with their customers and introduce new sales, marketing and delivery mechanisms.

Even as far back as 1498, the publisher Aldus Manutius shared his up-coming printed publications with his customers through a catalogue.[23] Later examples include the first shipments using the post as a form of delivery. William Lucas was a pioneer of this as early as 1667. He sent seeds to his customers. Then there's Pryce Pryce-Jones, who was acknowledged as the inventor of the mail order system - what a great name for a mail order king — nominative determinism at its best! His success came in part as a result of two technologies, the creation of the Uniform Penny Post in 1840 and the rail network extending to his store in Newtown, Wales.

More recent technologies were quickly adapted for purchasing across new mediums. Some, however, were a little more accidental. Home shopping via the telephone, for example, was the consequence of a small radio show in 1977 being given 112 can-openers by a customer who couldn't pay the advertising bill. The radio station owner, Bud Paxson, wanted to get his money out of these unwanted can-openers, and decided to sell the items live on air. [24] To the surprise of the radio host Bob Circosta, and Paxson, the items sold within a hour. Seeing the enormous benefits in this approach, Paxson later set up the world's first shopping channel on cable TV, which later became Home Shopping Network. Today it's worth more than two billion dollars.[25]

The internet was also a player in speeding up consumption. Although it had been around for a while, it was still fairly niche. Someone needed to create a more appealing way to use it, and most important profit from it. And that someone turned out to be Tim Berners Lee[26]. At the end of the 1980s Berners Lee was working at CERN, the large particle physics research laboratory near Geneva in Switzerland, home to some of the world's finest research scientists. The scientists had a problem sharing their findings and knowledge thanks to multiple different proto-internet-like comms platforms and unique systems. Berners Lee set to work creating a sharing platform for the frustrated scientists, developing three fundamental technologies which are still cornerstones of World Wide Web technology. HTML, or to give it its full name HyperText Markup Language, is the formatting language of the web. URL, aka Uniform Resource Identifier, is a kind of unique address used to identify each resource on the web, usually called a URL. And HTTP, (Hypertext Transfer Protocol) is how we retrieve linked resources across the web.[27]

The civilising of the internet in the 'www' form created by Tim Berners Lee naturally led to it being used as a new shopping channel. The pioneers of this new medium are now some of the world's biggest retailers. Online sales now account for more than one third of total sales in the US - $341 billion in 2015, which is up from $298 the previous year. [28]One of the most successful and earliest companies to utilise the internet as a shopping medium was Amazon who, in a later desire to speed up consumer purchases, created 1-Click shopping.

Payment couldn't be made more quickly than '1-Click'. In fact the technology goes well beyond the button. What is key to the idea is that the consumer's banking details are stored securely on servers, which means a returning customer can purchase items online instantly. Once Amazon had the patent granted in America, it quickly became a quality standard for online purchases. They never got European approval for it, despite repeated efforts which were rejected again in 2011. This hasn't deterred them from licensing the software to the world's biggest

www.amazon.com/Amazon-Prime-Air

brands. Amazon are now well on the way to pioneering a new style of physical delivery as well, via Amazon Prime Air, a service aiming to deliver packages of less than 55 pounds within 30 minutes via a drone flying at 400 feet to the customer's garden and landing on an Amazon marker (people in apartments are not a targeted customer for this I guess).

As a consumer experience it's pretty exciting. Imagine the anticipation as you wait for a drone to appear in the distance, watching from your bedroom window as it slowly gets closer and closer, then deposits your purchase in the garden. In fact it's easy to imagine the consumer lifecycle becoming almost meaningless in the wake of such exciting purchase and delivery experiences. Who would care for packaging, first time use, or quality of materials, when the delivery is so exciting?

Identity to currency

Currency is changing. Cash is disappearing. Sweden is pioneering this - and I am loving it as a recent migrant to Sweden. I managed to not use cash for a full four months when we first moved, relying instead on cards and apps. A recent article in the *New York Times* highlights the move, stating *"bills and coins now represent just 2 percent of Sweden's economy, compared with 7.7 percent in the United States and 10 percent in the Euro area."*[29]

Cash provides a certain anonymity. Going all digital, whether via cards, apps, or Contactless card payments, will link every purchase back to the consumer's identity. Jacob de Geer, a founder of iZettle, which makes a mobile-powered card reader, points out that *"Big Brother can watch exactly what you're doing if you purchase things only electronically,"*.

This jigsaws well with changes in the way we buy digital products. Many start-ups and digital companies prefer acknowledged attention via your profile over cold, hard cash. Eyes on the website, likes on Facebook, followers on Twitter or 'daily active users' all are accepted forms of payment to the digital product company. This seems like a good deal. Which it is if we are only interested in on-boarding, are happy to overlook long term identity usage by a company and forget any hope of comfortably off-boarding.

Quick and tethered

Our past as consumers was recognised via what we used and possessed. This was too slow for our economies and factories, so redundancy was introduced to increase turnover of consumed products. This benefits factories and businesses, and stimulates our egos emotionally, as we show off this year's brand new car model or the latest fashionable trend in kitchens.

The painstaking process of saving up to buy these goods was removed thanks to new types of credit and debt. As they evolved, cash lost its allure and sophisticated credit accounts became king. These methods of payment have matured and the systems we use have digitally tethered our identity to every purchase. The consumer is no longer just someone who buys something with his or her money. They give manufacturers much information about their identities in return for products and digital experiences.

These approaches do little or nothing to encourage reflection and responsibility in the consumer. So much emphasis is placed on the route to repurchase or engagement between the consumer and supplier that we overlook the off-boarding side of the consumer life-cycle. We seem to believe it to be of little worth in the value of the product, as a role in the economy, or as a message for the consumer, but this has left us exposed in a world of changing global temperatures, personal bankruptcies and fumbling digital privacy. We need to re-visit the overlooked area of endings, closure experiences and off-boarding. This is not to say that we need to stop buying things, damage the economy, or slow the rate of innovation down, but we really do need to think again and revive an interest in a wider, more wholesome customer lifecycle.

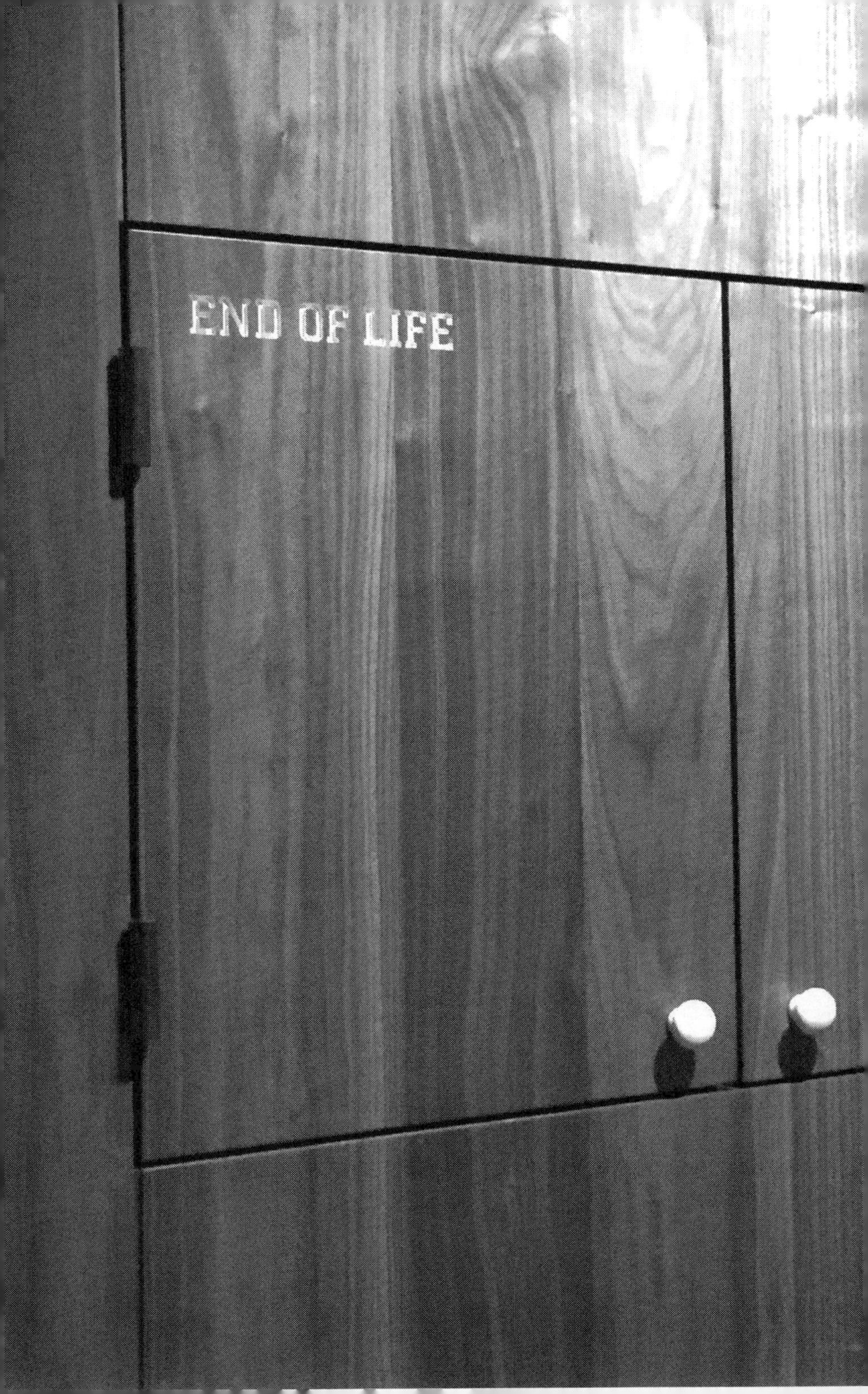
END OF LIFE

Chapter 5

The psychology of endings

Endings have a potential to fascinate humans, and the discipline of psychology reveals the variety of theories around those endings. Some psychologists say we're repelled by endings on a deep societal level. Others propose that resolution in life is important to the human psyche. Still more believe the end of an experience is a vital part of the memory-creating process. One thing is certain - we need to understand more about the off-boarding or ending part of the consumer experience to reduce its negative effects. In this chapter we'll examine the theories and look at how they influence closure experiences, whether they support the feelings that help us to self reflect, consider responsibility or acknowledge the change between something that is alive and something which is dead? These psychological triggers also influence different phases of the customer life-cycle. The on-boarding phase - which takes the customer from ignorant consumer to active customer - has been the target of many a psychological investigation in the drive for better sales conversion rates. Stimulating deep human emotions with stories that satisfy our higher cognitive needs of self actualisation is potent stuff, and businesses harness it to attract new customers. The usage phase of the customer life-cycle, when we are actually using the product or service, has also been the subject of investigations that observed shortcomings in products and services and gleaned insights that brought about improvements, innovation and product development. These in turn make products more usable and safe. But the off-boarding phase is usually neglected and ignored by the consumer industry. What little work is done takes place at the behest of consumer groups, governments and their quangos.

Commerce wants us to continuously buy and use new products, but overlook endings in these sometimes complex relationships. If they used similar sophisticated techniques popular at the on-boarding phase of the customer life-cycle during the off-boarding phase, we might access a more powerful and actionable approach to the issues that plague consumption. We might understand the reasons we avoid endings, both as consumers and providers. We might even improve the negative effects of consumption more effectively, efficiently, and comfortably.

While there is a lack of specific investigations into the end of the customer life-cycle, there's a wealth of relevant research into other types of ending. Our morbid fascination with our own death is one of them, and we can also draw on work in the memory field. Memory is important for understanding endings and influences other areas of human psychology. Our investigation will touch on some of these, too.

The basics - memory, recall and predictions

Memories are captured through the senses thanks to a working memory that assembles meaning in what we see, hear, feel and smell. The brain disposes of any distractions, rather like digital file compression, and the process converts the short term memory into a long term memory. According to the psychologist Endel Tulving, memories are captured in two ways, via episodic memory and semantic memory.[1] Episodic memory is a more personal and emotional memory. It captures events that you experienced, things like the people who were there and the emotions you felt at that time. [2]Semantic memory remembers common facts that don't come from personal experience, for example the names of colours, capitals of countries, and the spelling of words - facts you've collected through a lifetime, the sort of stuff we can all recall from school.

How we remember an experience, even a commercial one, impacts on the feelings we have about it happening again. Do you remember the event fondly, or with revulsion? Do you want to return to that beach or go somewhere else? Our commercial experiences often come with episodic and semantic content. The price and features of a new laptop would be a semantic memory, but the associated episodic memories you have about the device influences your feelings about it. I associate IBM laptops with the tedious corporate tasks I had to deal with when working as a manager for Nokia. My laptop was perfectly functional, but I had zero emotional interest in it beyond its cold utility. My Apple computer, on the other hand, was actually a dog of a corporate machine. It didn't sync up with any of the corporate back-end software I used, but because it was such a joy to use and Macs were loved by everyone in the design team, I had a strong emotional attachment to it.

Connecting associations with positive physiological triggers is the basis of branding. Obviously we can't have memories of something we haven't experienced, so telling stories through the medium of advertising helps create this kind of association. The sad part is that advertisers and marketeers have become so skilful at telling stories about ownership of a product, or usage of a service, that we've forgotten how to build good memories about the end of an experience. Describing the end of a product's life in positive terms and helping us to maximise good feelings associated with it would be a lot more positive.

Memories are also a key part of how we predict the future. We build on experiences to assemble and assess memories that help us picture and plan future events. This time travelling process uses similar parts of the brain to those we use to recall memories. It's a unique technique, found only in humans and therefore of significant benefit to our evolution.[3] Tulving called the capacity to mentally represent ourselves and imagine our protracted existence through time 'autonoetic consciousness'.[4] The very act of consumption uses every aspect of these memory skills. We time-travel to imagine the benefits a new purchase will bring. We remember how the purchases worked or didn't work in the past. We assess all this to establish how good or bad a deal is, projecting the consequences of a purchase using this knowledge. To make it happen, we need the experience to draw on. Without it we only have questions, which are no help in pinning the future down. Knowing how an event turned out not only requires insight to how it started and how went, but how it ended. A complete projection requires a complete experience. And we can't always have that. Predictions are a lot easier with hindsight. Without it we're no good at predicting. We get distracted by emotions.

Effective forecasting

First called hedonic forecasting by its originators Kahneman and Snell in 1990, this explored the accuracy of people's predictions. They found that we cloud the future with emotions from the present. For example, worrying about what to wear to a party might cloud your judgement of how much you're going to enjoy that party. An extreme but satisfying piece of research for those of us who haven't won the lottery was conducted by Philip Brickman. He interviewed lottery winners and registered their happiness levels before, during and after winning. He found that after only a couple of years, people returned to their previous level of unhappiness, unaffected by their new-found wealth.

Another interesting example, and maybe one many more of us will experience, is the issue of living wills. Gilbert and Wilson investigated people with a living will who stated that they didn't want medical intervention once their quality of life started to deteriorate.[5] However when investigating those with a low quality of life at the end of their lives, the researchers discovered that these people would go to extraordinary lengths to add just a few days to their lives.

When applied to consumer purchases, and particularly the on-boarding phase, Wood and Bettman found that people are more likely to overestimate, than underestimate, the duration of pleasure a purchase would bring. Advertisers and marketers encourage these feelings by creating wholly positive perceptions of the future thanks to a new purchase, images that project us into a happier, healthier, and more popular emotional landscape. We can see that the on-boarding phase is dominated by daily doses of positive advertising. Some say we're bombarded with 3000 adverts every day, many of which go unnoticed because they have to be familiar to us, or at least resonate, to be effective.

It comes as no surprise that adverts aim for short term emotional aspirations. Even mentioning the fact that a mobile phone will be outdated in 18 months would put many a customer off. Our purchases rely on overlooking long term facts like that. Instead we focus on the on-boarding and usage, remaining in denial that off-boarding is an issue, or that it should even exist. This isn't just an issue for consumers, providers and producers. As humans impact on more and more aspects of our planet, it becomes an issue for all of us, a critical problem we need to grapple with. And it's going to be one heck of a challenge, since we so rarely get to experience endings in the customer life-cycle.

The availability heuristic

People are drawn to familiar experiences and messages, and predicting the outcome of a new experience is much harder than making predictions about a familiar one. We find it easier to imagine experiences we've had before. Take sea level rises and oil spills. Few of us have experienced the impact of sea level rises. They're difficult to picture and hard to reference. It's much easier to imagine an oil spill because we've all seen them on TV and they happen with distressing regularity. We experience this availability heuristic when it comes to closure experiences, those avoided so often in our roles as consumers. People are eloquent about early parts of the consumer experience but lack the vocabulary around endings. How do we stop purchasing and consuming? The recent climate change debates in Paris and the many previous climate summits are a good example of how we find this subject difficult. On one hand we know we need to do something, on the other hand stopping what we're currently doing seems so hard.

On an individual level, when products or services break down, this usually means that a new version or an upgrade needs to be bought, or supplier changed. This kick-starts another on-boarding experience and dodges the off-boarding stage altogether. The end of the previous experience is brutally cut short without an opportunity for reflection, and the customer life-cycle rarely ends coherently. As a consumer, off-boarding isn't a nice memory to be recalled, and we don't think about the consequences of not doing it. An interesting example of our overlooking the end of the experience comes from the world of dieting. The author, physiologist and magician Paul McKenna puts forward four simple rules in his book that promises '*I can make you thin*'.

1. When you are hungry, eat
2. Eat what you want, not what you think you should
3. Eat consciously and enjoy every mouthful
4. When you think you are full, stop eating [6]

Far from being the usual austere approach to dieting, McKenna advocates consumption, telling the reader to enjoy what they want and when they want it. But, and this is the big thing, he asks them to reflect on the food they're consuming and acknowledge when they are satisfied. He reinforces the approach by saying, "*You can eat whatever you want, whenever you want, so long as you fully enjoy every single mouthful.*"

A short excerpt from his book highlights the theory behind the approach:

"*I've noticed a funny thing about people who are overweight. They spend all their time thinking about food — except when they're actually eating it.*

Then they go into a kind of eating trance, with a shovel as much food into their mouths as fast as they can without actually chewing or tasting anything.

Strange as it may seem there is a very good reason for that. When we do something that is essential for our survival, like eating, breathing deeply or making love, we release a 'happy chemical' in our brain is called serotonin.

People who are overweight often shovel food into themselves as quickly as possible in order to get a serotonin high. Unfortunately because they are eating unconsciously they never notice the signal from the stomach that lets them know they are full. So they keep on stuffing their faces, expanding their stomachs and putting weight on."

This could easily reflect our entire approach to consumption, and not just the way we eat. We're metaphorically shovelling consumption down our throats, uninterested in the taste and the aftermath. We can't stop eating entirely, we'd die without food – if we stopped consuming the economy would suffer. But we can consume with reflection, enjoy each experience, value it and when it's over, respect it.

Freud: the death instinct

Amongst Freud's less generally-accepted theories was the death drive; a self destructive compulsion towards death. Through his work on the pleasure principle, that suggested humans seek out pleasurable experiences, he came across some conflicting behaviours that made him question the logic of his principle. He found evidence of the death drive in three problem areas, that didn't adhere to his original theory. First of all, he saw that war victims would repeatedly return to traumatic experiences of war. They re-lived them time after time, which did not fit with the pleasure principle. Secondly he observed that in children's play, such as hide and seek, that when the parent disappeared, the child did not display the expected distress. Thirdly, he observed that his patients who had repressed painful experiences repeated those experiences as new versions of the old ones, instead of accepting that those experiences belonged in the past.[7]

The death drive was originally, quite a speculative idea, even according to Freud, who acknowledged that it might be far-fetched when he first theorised it. However he re-visited it often in following years, developing his argument to the point of saying that *"the death instinct would thus seem to express itself—though probably only in part—as an instinct of destruction directed against the external world"*. He developed this further over the years, until he came to the belief that *"libido has the task of making the destroying instinct innocuous, and it fulfils the task by diverting that instinct to a great extent outwards. The instinct is then called the destructive instinct, the instinct for mastery, or the will to power"*

It's fair to say that this aspect of his work never received the interest that other parts received. In regards to closure experiences it is of interest as a concept that may influence our thinking around endings or our emotional attachment to them. It is possible to attribute the desire for new, pleasurable experiences in the customer lifecycle to Freud's pleasure principle. At the off-boarding of the customer lifecycle we might well be seeking out some level of destruction. Maybe the added satisfaction of doing this against a horrible service provider or poorly made product, would be all the more pleasing.

Terror management theory

One of the most chilling psychological theories around closure is the idea that we're all trapped in a lifetime of denying death. The anthropologist Ernest Becker provides insights into the behaviours of society and our subconscious with his theory about our all-consuming denial of death. In his Pulitzer Prize winning work *The Denial of Death*, he argues that most human action is taken to avoid or ignore the inevitability of death. [8]We 're so terrified of it that we spend our entire lives trying to make sense of it. This drives societies to create cultural significance around life - religious beliefs, laws and symbols that explain the significance of our own version of living. We reward people who exemplify this culture and its values, and we punish or kill those who challenge it.

Ends.

Becker believes we have a propensity to invest in immortality projects, creating or becoming part of things designed to last forever. He suggests that the traditional systems we've put in place to resolve our questions about life are not relevant in a world of reason, leaving science to solve the problem of humanity. It's a pretty doom ridden exposé of our life on earth, which represents us as being reduced to desperate legacy creators. But it provides a rationale for our repulsion of endings, to the degree that we don't want to acknowledge them in something as simple and inanimate as consumer experiences. We just want to carry on consuming without thinking about the consequences, even if, as in the case of climate change, denial risks our lives and those of our children.

Research in 2000 by Kasser and Sheldon expanded that notion to suggest that humans pursue a materialistic life to deny death. [9]They asked two groups of people about their financial status over the next 15 years. The first group listened to music beforehand, the second was asked to think about death. The results revealed that the second group had a far higher expectation of their financial worth, and particularly higher expectations of what they might spend on luxury items. This suggested our concerns about mortality, although sub-conscious, strongly influence our behaviour and aspirations about material goods and economic status.

This is particularly relevant to closure experiences, supporting the notion that we deny closure in our consumer relationships in an attempt to cheat death by being 'in' a consumer experience, and by starting new ones. This makes sense – we don't want to end consumer experiences because endings remind us we're mortal. This is a strong motivation, and might account for our general inability to negotiate the end of any consumer action, whether it's a diet, climate change or banking restrictions.

In stark contrast to denying our ultimate ending, many of us actively seek a stable definition of endings when they happen. An ambiguous ending is worse than no ending, I guess, but we do seem surprisingly tolerant of bad and inconclusive endings. In 1996 the social psychologists Kruglanski and Webster posed the theory that we all need to experience cognitive closure in our lives. They suggested we have a definitive need to seek out information and find answers to ambiguous situations, improving our ability to predict the world with increasing accuracy. According to Arie Kruglanski we experience this in two ways. We seek closure quickly - the urgency tendency - and we want it to be maintained for as long as possible - the permanence tendency. This helps people 'seize' and 'freeze' when making judgements, which in turn reduces information processing. Different life situations demand different closures. Sometimes we want quick, firm conclusions to help us move on. When we place an image online, we're looking for the instant satisfaction of seeing that all-important picture of our cat or our dinner out there in the world for all to see, the urgency tendency. In contrast a pension requires a long term stability, unchanging over decades, and here we yearn for the permanence tendency.

Need For Closure Scale

Our need for cognitive closure is also affected by situations. In 1993 Arie Kruglanski, Donna Webster and Adena Klem created the *Need For Closure Scale* (NFCS) to help 'operationalise' their theory.

The scale refers to 42 situations that capture an individual's feelings about closure. People who score high prefer order, predictability and absolutes. Those who score low on the scale have a higher tolerance of ambiguity and openness, and are happier with interpretation. High scores on the scale correlate with conservatism, including social and political leanings, and low scores correlate with liberal creatives.

People seek all sorts of detailed information at the on-boarding phase of a purchase, things like calorie count, price, technical specs, sizes and sources. Once bought and used, there's little regard for how the item might be dismantled, broken down or turned off. While the producer might have made efforts to make sure their products are made of materials and chemicals that have these capabilities, the information is rarely extended to users.

In a product context, ownership has a great deal to do with possession. We established this type of ending over centuries, by physically walking away from products when we disposed of them, removing our association with the product and our personal responsibility for it. This was fine when products were relatively inert, but as products like consumer electronics get more complicated our usual way of ending product relationships looks too simplistic. The user needs to take responsibility seriously, so she or he can make good choices at the end of the product's life.

When it comes to services and digital products, by nature less tangible, the end of the customer life-cycle is less well defined. We want confirmation from service providers that the service has ended, usually in the form of a piece of paper or the end of a payment schedule or even a guarantee. That piece of paper provides us with the permanence tendency we crave. With digital products, things are nowhere near as clear cut. Removing the software from your PC or phone gets rid of it physically - or at least visually. But behind the scenes your name is still on a database and your data still being accessed. There's no tangible ending to the digital experience, and no permanency in its ending.

The dry, cold endings we have come to expect from banks, and other services that satisfy some level of permanency or urgency, are under-serving us. They have no emotional meaning. They're one-dimensional. And they fail to encourage the reflection and thoughtfulness we need in a fast-warming world, a place where billions in miss-sold finances and a steady erosion of personal privacy will haunt us for years.

Our experiencing self and our remembering self

Brands and manufacturers overlook the ending of the customer relationship as a way to build a good reputation. Most panic about customers leaving. The only thing this does is reveal the provider's ignorance. We already know that a big part of the way we recall an event concerns how it ended. Providing a poor ending is just as damaging as giving someone a terrible service. One theory strongly supports appropriate closure experiences, detailed in the work of psychologist Daniel Kaheman. In his book **Thinking Fast and Slow**[10] he reveals how we lay down memories in two ways, via our experiencing self and our remembering self. As he says:

'The experiencing self is the one that answers the question *"Does it hurt now?"* The remembering self is the one that answers the question *"How was it on the whole?"*. *"Memories are all we get to keep from our experience of living and the only perspective that we can adopt as we think about our lives is therefore that of the remembering self."* In this sense he suggests we remember experiences as *'peaks'* and *'endings'*.

These findings emerged from a piece

Thinking fast and Slow. Daniel Kahneman.

of research conducted in the early 1990s. It examined 154 patients who had had colonoscopy procedure, an uncomfortable but not harmful process lasting anything from four to sixty nine minutes. They were asked to assess their own pain levels every 60 seconds. Although Patient 1 and Patient 2 had different procedure lengths they both experienced a similar level of pain when asked afterwards. [11]Perhaps counter-intuitively, the experiment suggested the patient's pain wasn't connected to the length of the procedure.

Whether developing a new service, physical product or digital product, manufacturers focus on designing for the experiencing self, hoping that repeated good experiences will be recalled by our remembering self. This is a short sighted approach if we want to maximise the products positive impact.

Peak end rule

The second observation made by Kaheman was his peak end rule, where people judge experiences at two points rather than judging the totality of the experience: the peak, where there's an intense moment of experience, and the feeling at the end.The duration of the experience has no bearing on the experience as a whole. A quick glance online to investigate *"How to deliver a good customer experience."* brings up the usual recommendation, *"You must deliver a good service all the time and this will result in the end being really good for the user."*. These arguments focus on repeated peaks building up to a good ending, and our denial of real endings persists. We need to think carefully about the products and services we design, creating real endings, not mis-labelling the end of a task or sequence. In the context of a banking app, relating it to the peak end rule, users will remember a high-performing app function really well - *"I loved how simple it was to transfer money to another person."* and will also recall clearly what the end of the app was like, the reason they left the service, for example, *"I had to leave because we moved banks with our mortgage"*.

Too many companies overlook endings as a way to build reputation, believing customers should never leave and under-invest in closure experiences. It's no surprise people often have terrible experiences when leaving companies. If you believe anything about the peak end rule, you only have two shots at impressing a customer. Maybe investing in the end of the customer life-cycle would be money equally well spent. Sky provides a perfect example of this neglect. In 2015, displaying the usual paranoid business approach to customers leaving, they managed to damage their brand and give customers terrible memories. People who wanted to leave Sky were trapped in sales conversations for over an hour, while customer service team members tried to prevent them from leaving, sometimes even trying to sell them more Sky products. For a naturally polite person it was difficult to resist, and many people didn't leave, instead signing up for products they didn't want. It's a little like holding on to someone's leg when you get dumped, not allowing them to leave the relationship - it doesn't build respect.[12]

Concerned citizens and consumer groups suggested 'hacks' to get around Sky's system. Some recommend that customers lie, saying they were leaving the country. Others suggested people cancel their direct debits and write confirming their action to Sky instead of doing it over the phone. Some went as far as recommending people ask for a copy of the recorded phone call with Sky, a legal right under UK Data Protection law. What I find amazing about Sky's heavy-handed approach is their total misunderstanding of how closure experiences work. First, Sky were in denial that anyone would really want to leave them, second they refused to let people leave. It was rather like being kidnapped.

You and I know everything comes to an end, and that endings matter. Why do companies fail to make endings a positive experience? The Sky débâcle made sure that the experience of leaving was so awful that people recalled it over and above any positive experiences they may have had with Sky. If you only have two potential opportunities - the peak and the end - to give customers a good memory, it makes sense to make the best of both of them. Sky managed to ensure its customers had a dreadful ending experience, almost guaranteed to be a bad memory. You don't need to be a graduate in branding to know it was no way to build a loyal customer base.

Role Exit

When you understand the behaviour of someone who wants to leave, you have a powerful tool at your disposal. We all go through a similar sequence of thoughts when thinking about ending something. Knowing the steps people go through aids good closure experiences. In her book, *Becoming an Ex*, Helen Rose Fuchs Ebaugh investigates the transitions we experience when moving between roles.[13] In the past we had far fewer roles - one marriage, one job, one career, one faith. In modern society we tend to shift roles more frequently: serial career changes, a variety of jobs, multiple relationships. Ebaugh, herself an ex-nun, looked at a broad range of people who've moved from one role to another, including ex-convicts, ex-alcoholics, ex-cops and transsexuals.

Ends.

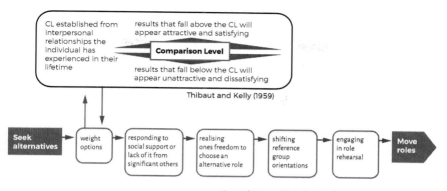

Thibaut and Kelly (1959)

Second Stage of Role Exit. Helen Rose Fuchs Ebaugh

First Stage of Role Exit

She found that in the first stage of role exit, doubts are often ignited thanks to organisational changes, personal burnout, changing relationships and the direct effects of events. These doubts are reflected to peers and friends in the form of cuing behaviour. These cues are recognised and reinforced by others, prompting further investigation into role changes and easing doubts. If they don't get reinforcement, people will re-evaluate their intentions. If someone is not given reassurance that there's a strong basis in their doubts, they put more areas under scrutiny, and subsequent events will be received as negative reinforcement.

Second Stage of Role Exit

Once they realise they need to change roles, they move to the second stage of role exit and start searching for viable alternatives. A key part of this is the comparison level, created by Thibaut and Kelly in 1959. This level is established via previous life experiences, which form a baseline to judge subsequent experiences against, either above the level - attractive and satisfying - or below it - unattractive and dissatisfying. After assessing the alternatives, the individual looks for reinforcement from significant others. If reinforcement is given, the individual feels a sense of freedom and confidence, and they'll begin practising their new roles in preparation for the potential new role.

In relation to the customer life-cycle, Ebaugh's work highlights events that might drive a person to leave a commercial relationship. Price comparison sites are a wonderful example, known for starting doubts in the consumer's mind. Many industries now have independent websites reflecting prices between competitors. Whether you're searching for better insurance, broadband or banking, they reflect commerce's inability to help people move freely between suppliers, which of course brings an end to the relationship with a supplier - a closure experience. Price comparison sites not only introduce doubts, just like the examples Ebaugh found, they also help the consumer move to a competing service. In reality the price comparison site provides a full and complete closure experience, ironic when at the same time price comparison site creators deny closure experiences even exist.

Whenever we engage in consumer activity, we weigh up the pros and cons. Is it too much, too big, not big enough, too fast, too much of an indulgence? And all this uncertainty has a long term social impact. Social dilemmas happen to us every day. They encourage self reflection and personal responsibility for purchases. They also require thinking about the long term impacts of a purchase after we've finished using it. Good closure experiences should encourage reflection about long term impacts.

Social dilemmas

The banking industry is awash with issues around social dilemmas, with plenty of examples of people making decisions that benefit themselves over society. The Libor (London Interbank Offered Rate) scandal in 2012 is just one of them, when a small group of individuals manipulated an exchange rate to their own ends.[14] It was a shock to the financial services industry because almost all of their lending is related to Libor, from student loans to financial derivatives, totalling an estimated 350$ trillion. Although we might criticise the people involved, we all take part in a version of social dilemmas every day, in small ways. Consumer product experiences come with risks. Take clothes shopping, where we might be attracted to a fantastic new top at an incredible price. The price might raise suspicions about poor labour practices. Perhaps the top is made by employees with no employment rights, who work in dangerous conditions. But we might push these thoughts aside, seduced by the short term pleasure of the purchase.

The same goes for jumping on a flight to take a holiday, an action that's loaded with perceived benefits to help consumers cast aside their social dilemma. Advertising messages encourage us to ignore social dilemmas like climate change in favour of short term gains like enjoying a foreign holiday. Air travel comes with notoriously high CO_2 emissions and the industry lags way behind other forms of transport in 'going green'. Concerns about long term damage to the environment can be neutralised or at least acknowledged through the planting of a tree, off-setting the carbon released as a result of the flight. [15]

George Monbiot, the author of **Heat**, suggested carbon offsetting just allows rich individuals to neutralise their guilt and carry on filling up their 4x4s without changing their behaviour. This, however, has been challenged by the offsetting industry and Britain's National Consumer Council, who argued that many people use off-setting schemes as a part of a wider carbon reduction strategy, stating that *"a positive approach to offsetting could have public resonance well beyond the CO2 offsetting, and would help to build awareness of the need for other measures."*

Carbon offsetting is a vital part of the customer experience if we want to encourage personal reflection in consumers, and it's a fantastic example of closure. Monbiot's opinion is clouded by social issues and ignores improving the consumer experience and rebalancing it towards responsibility. We have to engage in the customer life-cycle to change behaviour. Closure experiences attempt to make that change from within. We can't judge consumption from afar. We need to yield its tools, to improve the process. The real issue with carbon offsetting is that once again we've talked ourselves out of exposing consumers to responsibility. Airlines hide the issue of carbon deep in their websites under the heading, 'social responsibility'. The irony is, this removes that very responsibility from the people who should be responsible - consumers. A few years ago it was an issue everyone had to recognise when buying a flight. Now we can skip the closure experience of air flight.

We experience a wide range of psychological behaviours throughout the customer life-cycle, and they influence us every day. I've mentioned a handful that are relevant to closure experiences, and they all manipulate the psychological urge we have to consume. These urges are encouraged by the consumer society, which helps us to indulge ourselves in modern life and distract ourselves from the inevitability of death. Challenging these urges - and the status quo of Western society, business and our individual consumer demons - is no small feat. We need to use the numerous skills we've developed in building starting experiences that motivate the customer life-cycle towards more purchasing, and re-deploy those skills to develop great closure experiences at the end of the product life-cycle. This way we can engage active consumers in self-reflection, supporting the fulfilment of responsibility in a meaningful and universally beneficial way for both society and business.

Chapter 6

The narrative of endings

We can all be inspired by the way literature and film deal with narrative and closure: a wealth of knowledge, spanning centuries of creating meaning, laid out in a sequence of events that tells a story. But most of our product and service stories fade out before we get anywhere near a clear ending. Commerce is overly focused on the starting and usage phases of the consumer experience, overlooking the ending as a useful and meaningful moment.

Endings are important. They establish important fundamentals for ourselves and wider society. Richard Neupert references Walter Benjamin, Sartre, Hayden White and Peter Brooks as writers who all believed strongly in the benefits of closure in the narrative. In his book *The End, Narration and Closure in Film* Neupert says that *"solid closure in conventional narratives and histories satisfies individual and social desire for moral authority, a purposeful interpretation of life, and genuine stability"* [1] In parallel, Elizabeth MacArthur in her book *Extravagant Narratives* calls it an *"attempt to preserve the moral and social order which would be threatened by endlessly erring narratives."* [2]

We could easily describe many of our man-made consumer

experiences as having 'endlessly erring narratives'. Adding coherent closure experiences would add 'purposeful interpretation' to our consumer endings. It might also improve our 'social desire for moral authority' which, let's face it, is lacking when we discuss the negative consequences of consumption.

Despite recognising the enormous, world-changing issues we face as a direct result of consumption – things like climate change, peak oil and the miss-selling of financial services - we appear totally unable to come to a conclusion about vital issues. Some see this as an example of a lack of moral authority brought upon us, thanks to bad or absent endings.

Adding endings to beginnings

There is a deep rooted history of narrative structure going back many centuries. It shows clearly how moments in the narrative need to relate to one another - a start relating to an end. Aristotle observed how his contemporaries' work often had a broad three act structure. Act 1 develops the wider context, creating a setting and establishing the characters. Act 2 develops tension and reveals the conflict in the story, also a common point where the love-interest turns up. Act 3 brings the tension and conflict to a climax, resolving the story with a conclusion.[3] Our consumer narratives reflect a similar path for the first two acts. Act 1, on-boarding, describes the wider context of the product, its features and consumer benefits. Act 2, Usage, gets to the action, revealing the reality of the product and establishing product loyalty. Act 3 lingers unwritten in most consumer relationships, left to another narrator to tell, a role usually taken by councils, governments and wider society.

Aristotle's three act structure was later widened to five acts by the Roman theorist Horace from 65-68 B.C. His adjustments were pretty successful and lasted for hundreds of years. Plenty of Shakespearean plays, for example, followed his five act structure, including *Romeo and Juliet*, *As You Like It* and *Macbeth*.[4]

Ends.

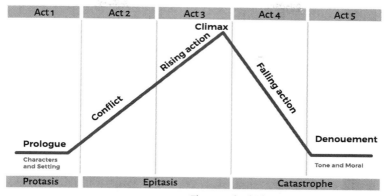

The 5 act play, laid over Aristotle's 3 act version

Roll forward another few centuries and the 19th century novelist Gustav Freytag saw narrative as a 'dramatic arc' involving exposition, rising action, climax, falling action, and finally the dénouement, the French word for 'untie'.[5] This structure sees the story resolved, conflicts settled, a sense of normality restored, a new equilibrium established and closure achieved. I love the phrase 'untie' as a source of conclusion in stories. It dovetails beautifully with the consumer experience.[6] It feels like there's more to the issue of ending than simply a halt to the story. The difference between endings and conclusions, according to Barbara Herrstein Smith in her book **Poetic Closure** is that, *"any event, narrative or otherwise, may simply stop or end; only a text or art work may conclude, with conclusion coming at a definite termination point."*[7]

We know a natural ending can happen to anything, an abstract, sometimes random event that's usually out of our control. But human-constructed endings - those in films, books and consumer experiences — can and should be thought through. The endings we don't currently design into products, services and digital products could be made meaningful rather than fizzling out without really entering our consciousness. The film industry uses a variety of tools to bracket our experience of their products, constructing real meaning for the audience. For example music, titles, the style of voice, the type of shot and lighting can all open a film and close a film satisfactorily. In Newport's words, *"Closure most often involves a stylistic framing of the text and the story; it may frame the text with parallel motifs, bracket the narrative by imposing similar opening and closing elements, return to the primary narrator, and attach the discursive closure devices to call an end to the narration."*[8] In the consumer experience we should make use of the same tools, providing the same emotions invoked at on-boarding and applying them at the off-boarding stage. There's a lot to be said for bracketing the consumer experience with a start and finish, using the same language, the same style and the same level of authority.

The consumer experience is often started off by a commercial entity with persuasive language, slick presentations and powerful promises. Think TV adverts! Sadly the experience is ended by a different entity, usually a municipal company or representative of society, using a different type of language, and a style that's often cold and full of blame. It really isn't the same kind of experience. If we improved closure experiences in the same way for the customer life-cycle, advertisers would encourage us to recycle appropriately, using the same authoritative voice they used to start the relationship instead of the current municipal council recommendations and instructions. We would listen to authoritative messages about why it's important to receive a pension from a provider, not just the importance of starting and paying into one. Putting the responsibility on providers, sellers and re-sellers to phrase, create and execute closure experiences as good as starting experiences might encourage more holistic thinking.

Talking to death

The narrative voice in cinema is a key authority, guiding us through a sequence of events and supporting a story's effective opening and closing. This voice has an important role to play. Gerard

Genette breaks this role down into three key responsibilities in his essay Narrative Discourse. One, tense, or story time relates to discourse time. Two, aspect, the way the story is perceived by the narrative voice. And three, the mood, in other words the type of discursive representation used. In contrast consumer voices are broken into different narrators.[9] Advertising, marketing and branding will often talk of the brand voice as part of the brand presence. The strategic marketing and brand firm Larsen puts it like this *"Brand voice is the purposeful, consistent expression of a brand through words and prose styles that engage and motivate."*[10] At the end of the consumer experience, the voice which describes replacement, or discard will be that of the civic role, of society. That voice will encourage responsible community actions like putting the bins out graciously, recycling properly, driving a high powered car slowly through a village, quitting smoking, and a variety of other social requests. The end of the customer life-cycle isn't the selfish indulgence we're fed at the beginning of the transaction. It is a group responsibility, and so today's endings are often group experiences. At the end of our lives people gather to share the experience of our passing. And at the end of a single life, the rebirth into a shared life through marriage is also witnessed by friends and family. According to Marianna Torgovnick, author of **Closure in the Novel**, the things we tend to expect from narrative endings are big communal ritualised events like weddings, births or funerals. These bring us together in real life, a transitional stage after which we separate and move on.

Modern stories

New formats and technologies have complicated the narrative structure, bringing long standing rules and systems into discussion. Some challenges have a similar source – endings. The TV show format has had a host of new problems to overcome. How to end a show after 100 episodes? It's a big question. Some shows have been guilty of going on too long, avoiding an ending despite being way past their best.

In the industry it's called 'Jumping the Shark'. Most shows get axed because of falling audiences. Few have the luxury of choosing their own ending, and even if they do it doesn't always go smoothly. In mid-2004, *Friends*, one of TV's most successful ever shows, ran for the last time after two hundred episodes. It had become so expensive that the network felt it couldn't justify paying for more. Plus, some of the actors, writers and staff involved were keen to move on. Everyone who worked on the show was emotionally prepared for an ending. This bled through into the final episode, leaving viewers feeling things were not resolved at all well, with poorly-crafted knee jerk conclusions to stories. It had 'jumped the shark' because no one could bear to put it out of its misery.

Dexter's eighth series run concludes with him mercy-killing his own sister, taking her body out to sea and ceremoniously placing it with the other people he had slain. However a coming hurricane smashes his boat and leaves him swimming for his life - which, had he died, could have made a conclusive and dramatic ending. But the executives at the network refused to kill off an asset that had been so successful and demanded that the scriptwriters write him into the final scene alive. This denied the audience an appropriate ending in much the same way we're denied proper endings as consumers. It's clear businesses of every kind, even TV, find it really hard to acknowledge that endings actually happen.

Friday Night Lights, which wasn't as big as *Dexter* but had many loyal followers, did a much better job of concluding lots of diverse storylines well. *Rolling Stone* magazine described the finale as, *"doing what it should do - wrap up a season's worth of storylines and give us a good idea of what the future might hold for the teens and teachers of Dillon, TX. In the end, hard choices get made but life goes on, and viewers feel like they're better for having spent time in the company of a small-town football team taking it one game at a time."*[11]

Some TV shows adopt a fanciful route. The final show of *St Elsewhere* suggested that the whole show had unfolded in the mind of an autistic child. And the last episode of *Roseanne* revealed that the whole show had been fiction written by the leading lady, while her husband had been dead all along. Although comical and often in keeping with the show, mocking the belief that the audience has committed to it. This is perhaps an example of our inability to deal with endings well in general. In cases like *Dexter* and *Friends*, it proved hard to end the shows because of their commercial potential, not because of the audience's deep need to see more. Humans can and do adapt to change and death. But maybe network executives and businesses have more of a challenge, mired as they are in denial.

Game over

The games industry also challenges the narrative model, displaying a complex array of moving targets and adopting a great deal from film and books. On a technical level games have an almost blinding quantity of opportunities - VR, faster internet speeds, networked players, increasing processor speeds, increasing sensor usage - to name but a few of the Pandora's box of tools available. The player is the action-taker rather than a passive audience member, and it's a similar role to the one we adopt in our consumer experiences. Amongst their biggest challenges is resolving the end of the game. Where once we'd expect to get to the end of the game, slaying a 'big boss' on the way, we now see a far more complex world for the gamer to pursue. We get multiple narratives, sub plots, sub games, and what seems like endless worlds to explore. The issue of an ending is changing, and this brings up lots of questions.

I raised this discussion with Ken Wong, a veteran gamer, an old colleague of mine and the lead designer for the multi-award winning game *Monument Valley*. It's installed on more than ten million devices and has made $5.5 millions for the company we worked for. Ken was always an insightful individual, and I was keen to ask him about his thoughts on endings and closure in games. We discussed the broader life experiences of endings, and with that death came up. He shared how he had been influenced strongly by the death of his mother. It was a profound and scarring incident for anyone, but it inspired the ten year old Ken's life with an appetite to live. He established the encouraging view that, *"asking about death, is then about life"*, something that drove his approach to travel, taking on challenges and embracing adventures, all part of a determination to live through endings and move on.

In Ken's words, *"Inevitably we say goodbye to jobs, people and places. These are temporary and fleeting sometimes. It's like concluding a chapter in a book. Games can vary in this regard, too. Some games you complete, other games don't have an ending. And some are constructed to provide both continuity and endings. In-game rewards can stretch the player to achieve a short term conclusion, yet provide an endless journey in the wider game. With Monument Valley, the ending became an aesthetic or philosophical question. We could have kept on inventing levels based on the theme, but we risked repetition. Instead we choose to make a finite length of gameplay, which in turn let us place more emphasis on quality. ...*

Games traditionally value hours of play as a measure. Boxed games reinforce this approach, with the significant investment of $40 from the player, but we're starting to see the emergence of a wider variety of games and platforms to create and play in. The increase in mobile and downloadable games is a good example. I remember watching a TV show about the future of consoles, with Sony proudly announcing that the PlayStation 4 wouldn't have a CD drive. At the time I thought this was crazy, but it makes total sense now. It gives the game provider a significant control over piracy, features and connectivity. And it provides the freedom to expand the game incrementally.

Some gaming companies have experienced a kick-back from the gaming community because of the need for a connection. SimCity is a single player game the majority of the time, and frustrated players who didn't see why they needed to be connected to the internet to play. On a wider, more philosophical note, I think we lose something when we constantly update online games - the original game is lost. I was recently asked by a museum if they could place

Monument Valley in their collection. But the effort to keep a digital game working - I mean proper archiving - is enormous. Every time Apple or Google updates their software, we have to update the game. Some games are left behind in digital history as a result. We can never enjoy that 'completed work' moment so common in books and films. For example, how would you play the original World of Warcraft game, launched in 2004? You can't, it's long lost in updates. As an audience, we have an enormous variety of games. We are spoilt for choice, and this leads to some strange behaviour, almost Stockholm Syndrome-like. We might commit to hours of game play yet find little valid conclusion or wider benefit for life. As a game creator this leads to diverse feedback from the audience. Some will celebrate the game, but others will criticise it despite playing it for hours. PacMan is a good example of endings. At the end the game crashes. It basically shouldn't be played to level 256. If you get that far, you're playing too long. I think a good comparison for games is like a theme park, you wouldn't, or maybe shouldn't just go on one ride all day. The best way to enjoy a theme park is to go on a ride, meet with your friends, go on another different ride, hang out, eat some food, and then another ride it's about experiencing a variety of games."

> "Having played over 400 hours of this game I can safely say you can avoid getting into it now"
>
> — 442.7 hours played

The blog Steam Review Watch illustrates the issue with endings nicely.
steamreviewwatch.tumblr.com

Game creators like Ken are challenging the established rules of narrative as the industry develops its offering. Starting as a simple boxed game for $40, the industry now creates immersive, long term experiences, and players are changing their attitudes too. According to a recent presentation by Tom Abernathy and Richard Rouse from

Riot Games and Microsoft Game Studios, fewer gamers these days are completing games, even those with a strong, compelling story narrative and a coherent closure.[12]

These are the proportions of gamers who, research found, were completing their games.

The Walking Dead: Season 1, Episode 1 — 66%
Mass Effect 2 — 56%
BioShock Infinite — 53%
Batman: Arkham City — 47%
Portal — 47%
Mass Effect 3 — 42%
The Walking Dead: Season 1, Episode 5 — 39%
The Elder Scrolls V: Skyrim — 32%
Borderlands 2 - 30%
Source: Steam Achievement Data / Bioware[13]

Tom and Richard believe this phenomenon is down to a change in attitude, with people not experiencing the game as an established narrative format any more. They believe that the three or five act structure built in games by many video game developers is redundant. One piece of evidence they put forward to support this notion is work done by Deborah Hendersen, a user researcher from Microsoft. She found, when asking players about the plot of their favourite game, that they were unable to describe it at length, yet they were perfectly able to describe the plots of movies and TV shows they liked. Personally I wonder if this is related to the way we experience TV and movies, as a single narrative allowing for comfortable memorisation and easy discussion. Gaming, in contrast, has a diverse plot structure with multiple options embedded within the wider three or five act structure. But it's a compelling finding all the same.

Many gamers, although not able to speak at length about it, find the ending a powerful aspect of the game. A recent article called the '10 video game endings that left you totally speechless' champions an alternative view to that of Deborah Hendersen. One of the games that came up in the top 10 was Spec Ops: The Line. It takes players on the usual first person shooter experience - achieving tasks, killing people, creating explosions, smashing stuff up - but there are other, subtler options along the way, and they reveal moral questions. This isn't entirely apparent until the end of the game, when the mental health of the protagonist is revealed, suggesting he or she has created a series of delusions to justify his or her actions and now needs to reflect on their behaviour. This, as you can imagine, creates strong emotions, and to certain degree makes gamers question the morality of the actions players take within the game. It's an approach that would be a useful addition to the consumer life-cycle, in which we thoughtlessly play the role of protagonist in a first-person shooter called Consuming, but fail to get to any meaningful resolution about why we're doing it. Creating this sort of reflection for the ending of an experience is a welcome approach to closure. We could learn a great deal from games like Spec Ops: The Line, which introduces powerful emotions that encourage people to reflect on their previous actions. And it certainly fulfils Elizabeth MacArthur's interpretation of endings as an *attempt to preserve the moral and social order"*

That Dragon Cancer

One of the industry's most challenging and subsequently most emotional games at the moment is That Dragon Cancer. Few of us would want to experience, let alone pay for, the emotions that the game generates. This is no ordinary game. It reflects on the real world frustrations, injustice and helplessness of cancer sufferers and their families. [14]At first glance it might not seem like a compelling subject to play at. But it has received amazing reviews and highly emotional responses from players.

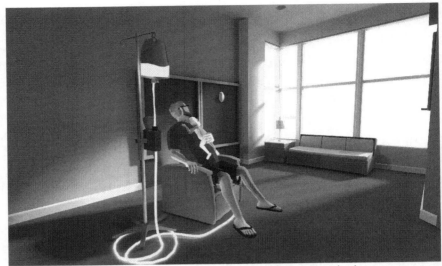

www.thatdragoncancer.com

The game's designer, Ryan Green, created the game to vent the emotions he and his family had been experiencing as their four year old son, Joel, battled with brain cancer: the trips to hospital, intravenous drips, sickness, steroids and the roller coaster of small wins, the cancer shrinking and then growing. While Ryan was in hospital one night with Joel, trying helplessly to soothe him, feeling totally frustrated with the usual dead end solutions, he realised his experience was like a, *"game where the mechanics are subverted and don't work."* [15] Ryan and his co-designer Josh Larson started to work the scenario into a game, translating a desperate parent's choices and experiences into game mechanics. The big difference is that the choices and experiences involved are nowhere near as simplistic as in an ordinary game. There's no easy, single solution. It's raw and complicated. It's hopeless and desperate.

Once the pair developed a demo, they started taking it to game conferences. It caused a lot of discussion and received wide acclaim. The game's journalist Jenn Frank wrote an emotional celebration after it triggered memories about her mother's death, and saw the role of the game as *"sustaining the hope and joy of life for just as long as we can"*. [16]

Ends.

That Dragon Cancer flips our usual expectations around endings in games. Where we usually experience a conclusion, or achieve a higher life, heaven or afterlife through being superhuman, using tools and achieving tasks, this game is very different. The player doesn't have simple choices. The achievements are not clear win. Endings are equally unclear. And people love it. Is there something important to be learned here?

Games ask questions of us. They challenge our perception of history and identity. But it's rare for a game to grapple with real life in such an emotionally engaging way. The idea of death tends to be placed flippantly in front of the player, cheapening the relevance of the player's reference to endings as characters are reborn again and again to keep the player engaged. That Dragon Cancer values precious life, holding on to it as it dangles on a thread. The player has no choice but to acknowledge life's importance. It's a real-life game and reveals how hard both conclusion and achievement are. It values endings, not because it presents them clearly through a narrative structure in a controlled format, but because it makes a comment on the real challenges in life in a real way. What it teaches us is this: death comes like it or not. Maybe our consumer experiences need to recognise this too, inviting us to appreciate the real consequences of endless consumption.

The tales we tell ourselves through narratives reflect meaning from the real world. The reason we watch, play and read these narratives is because of that real world influence, constructed in a meaningful way and delivered through the characters we relate to. Poorly-ended stories, whether in books, films or games, don't get an audience. In contrast we're sold countless broken stories in our consumer relationships. Stories that might last years, or even decades before they end. Tales that leave us lingering without a clear conclusion, undermining ourselves and society.

Ends.

Learning to tell completed stories in consumer relationships, with clear endings, will provide healthier societal and personal outcomes. It will support the bigger stories that the human race is telling, things like what we've done to the earth, our fellow creatures, the climate that sustains us and one another. When it happens we will finally enjoy the satisfaction of a purposeful interpretation of life. It might even deliver the genuine stability that Richard Neupert saw when he said *"solid closure in conventional narratives and histories satisfies individual and social desire for moral authority, a purposeful interpretation of life, and genuine stability"*.

Chapter 7

Love and marriage ...for life?

The largest-scale service we engage with throughout our lives is the government. Our taxes pay for it so as to finance the enormous range and levels of service delivery necessary to operate an entire nation. These services are intended to even out the advantages and disadvantages experienced by the people in the government's care. In theory, the more able help the less able. Young people are educated, the frail and the needy supported, our borders are protected, the police maintain law and order. From birth to death, there is no escaping the services provided by the state. Some of the very important messages that the government sends out about their services relate to marriage.

Love's social cul-de-sac

Although we can't define love as a service, we can treat its placement in society as a service. Marriage has many of the service characteristics that we see elsewhere in society. There's a powerful social story about its benefits, with family elders often encouraging younger people to tie the knot. Governments encourage it through tax breaks for married couples. The groups that define and provide marriage as a service are a mix of state, community and religious authorities, depending on your background, location and beliefs. Within the narrative we are told about the benefits of marriage, but we are also warned about ending a marriage.

As a child of the '70s, it wasn't statistically surprising that my parents divorced. UK divorce rates doubled between 1970 (58,239) and 1980 (148,301) [1] - the year my parents ended their marriage. A great deal of social change was happening at the time. The agreements many couples made at the altar, in front of their friends and family, were not carried out in practice. Luckily my parents ended their marriage fairly amicably. Many couples didn't. Plenty ended up in court, not speaking for years, which put enormous pressure on their kids. It was not uncommon for divorced parents to see each other for the first time post-divorce at their child's wedding twenty or more years later, when frosty additional pressure was added to the wedding day thanks to the previous generations' lingering relationship issues.

On the other hand we can't blame those who have had difficult divorces. The taboo associated with ending a marriage is enormous. Still to this day, society burdens divorced people with stigma, more so in some countries than others. It systematically outcasts them, suggesting they have made a moral mistake akin to a crime. Friends are often forced to 'pick sides'. Society turns its back on you, refusing to talk about the ending of the marriage it was so keen for you to undertaken. Much of this could be avoided if we had a better, more grown up approach to the end of a marriage.

Marriage is honoured almost universally by religions, which praise the bond between two people on their way to setting up a home and potentially starting a family. It was probably a good thing in the past, with a couple promising to look after one another and care for each other for the rest of their lives. High expectations were placed on marriage, considered the start of real life, not just the bonding of a couple. It was the start of living away from parents, a new house, a sex life, children, increased respect from elders and so on. The positive stories we associated with marriage were endless, a heavily biased on-boarding which represented a state of normality.

Marriage wasn't always a religious institution, however. Previously, couples were empowered to get married and divorced at their own will.[2] There was no need for religion, ceremonies, or officials. It was mainly a private process, according to Shannon McSheffrey, author of *Marriage, Sex, and Civic Culture in Late Medieval London*. Marriages were commonly conducted by the couple through mutual consent, by simply declaring "*I marry you*". This "*verbum*" was unquestionably binding, unless the marriage had not been consummated.[3]

Religion became interested in marriage once it realised the ceremony's true social potential. The Bishop Ignatius of Antioch wrote to Bishop Polycarp of Smyrna around the year 110, highlighting the opportunities marriage offered. "*It becomes both men and women who marry, to form their union with the approval of the bishop, that their marriage may be according to God, and not after their own lust.*"[4] But it took many centuries for Christianity to act decisively. Its patience finally ran out when the marriage of Diego de Zufia and Mari-Miguela in Zufia, Northern Spain, was denounced by Diego because it wasn't consummated. A lengthy legal battle resulted and eventually the marriage was annulled. Tired of the typically casual approach to marriages that were being created and then annulled, the Catholic church moved to change the law. In 1563 the Council of Trent required that a valid marriage be performed in front of a priest and before two witnesses.[5]

On-boarding bonds

The Catholic church, like most religions, holds strong views around the story of marriage for its 1.2 billion followers.[6] It tells a compelling narrative taken from the bible, providing clear guidance of the benefits of marriage in 'God's eyes' and encouraging its followers to pursue the accepted route to becoming good Catholics. The Vatican points out in The sacraments at the service of communion that, 'Man and woman were created for one another: It is not good that the man should be alone. The woman; flesh of his flesh, his equal, his nearest in all things, is given to him by God as a helpmate; she thus represents God from whom comes our help. Therefore a man leaves his father and his mother and cleaves to his wife, and they become one flesh. The Lord himself shows that this signifies an unbreakable union of their two lives by recalling what the plan of the Creator had been in the beginning. So they are no longer two, but one flesh.[7]

In most interpretations of the Hindu religion, the marriage can last across multiple lives with the concept of rebirth, as the couple progress spiritually. It is considered a 'relationship of the souls'. It comes as no surprise that the start of this multi-life bonding is a complex affair with an expensive dowry system, enormous weddings, and endless fine details to consider around caste and income, all of which are just as important as the emotional factors.[8]

Off-boarding bonds

Any bonding that involves people will naturally change over time as we grow as human beings. Experience of life will cause changes as our views and knowledge evolve. Marriages, regardless of religion, and the growing number of non-religious versions, should recognise this, but as with many services, the story told at the beginning may change by the end. The Catholic Church acknowledges the difficulties inherent in staying married for a lifetime, even to the point of saying in 'The Sacrament of Matrimony' that it is an almost impossible task. *"It can seem difficult, even impossible, to bind oneself for life to another human being. This makes it all the more important to proclaim the good news that God loves us with a definitive and irrevocable love, that married couples share in this love, that it supports and sustains them, and that by their own faithfulness they can be witnesses to God's faithful love. Spouses who with God's grace give this witness, often in very difficult conditions, deserve the gratitude and support of the ecclesiastical community."*[9]

Religions tend to vary in their views about the end of marriage. As a rule, it is not welcomed, especially as such heavyweight and important statements have been made 'on-behalf of', or 'in-front of' a deity. It would make it too easy to discard those statements. Despite the Catholic Church's admission that sometimes marriage becomes 'practically impossible', it still denies its followers the ability to formally end the marriage, insisting the institution is 'indissoluble':

"Yet there are some situations in which living together becomes practically impossible for a variety of reasons. In such cases the Church permits the physical separation of the couple and their living apart. The spouses do not cease to be husband and wife before God and so are not free to contract a new union. In this difficult situation, the best solution would be, if possible, reconciliation. The Christian community is called to help these persons live out their situation in a Christian manner and in fidelity to their marriage bond which remains indissoluble."[10]

Three Talaq

For many Indian Hindus there was no way out of an awful marriage until 1955, when the government created the Hindu Marriage Act and made it law. But even now women are still fighting for a basic level of status in marriage and divorce. Tenzing Chusang, from the Women's Rights Initiative, which helps women with legal support, says *"In India there's no such thing as shared matrimonial property or equal division of assets. All she gets if the husband divorces her, and that too after years of litigation, is a minimal maintenance payment. What can she do? She has to stay."*[11]

Far from ruling out divorce, it couldn't be quicker in some interpretations of the Islamic faith in India. Despite the majority of other Islamic nations banning it, a husband in parts of India can just utter 'Talaq' three times to complete a divorce.[12] It is not uncommon for this to be done via a text message, or even via Facebook, and it really does end the marriage immediately, with the wife expected to leave the family home straight away. Living in a marriage that comes with such a simplistic route to divorce leaves the wife at a horrifying disadvantage, threatened constantly by the prospect of an instant end to married life and terrified of being thrown out onto the street. Zaria Soman, an activist in India working for Bharatiya Muslim Mahila Andolan (BMMA), champions women's rights and campaigns for a ban on triple Talaq. She says *"It is a totally unilateral, one-sided, instant form of divorce, and uttered by men. The wife need not be present. She need not even be aware. According to a recent study by the BMMA, as many as 1 in 11 Muslim women were survivors of triple talaq — the vast majority receiving no compensation."*[13]

The centuries-old social narrative in many countries has long denied a reasonable ending to marriage. So it comes as no surprise that it is such an unwelcome thing to do. A person who ends their marriage is, to one degree or another, out cast by society. In many countries this powerful social disapproval falls unfairly and unjustly on women.

Some aspects of marriage, especially the institution's inability to tolerate endings, seem incompatible with modern life. It's scary when we think about the commitment we're making to one person for the duration of a lifetime. The comedian Aziz Ansari, makes light of this situation in a brilliant routine about the absurdity of expecting an infinite commitment.

"Imagine if marriage didn't exist?, and you're a guy asking a women to get married. Imagine what that conversation would be like..."
Guy: 'Hey, you know we been hanging out together for a long time?'
Girl: 'Yeah. I know'
Guy: 'I want to keep doing that until you're dead.'
Girl: 'Whaat???'
Guy: 'I want to keep hanging out together until one of us dies. Put this ring on your finger so people know we have an arrangement.'
Girl: 'Whaat?!'[14]

It seems the only way out of some marriages is to suffer until one of you dies. Religion isn't particularly interested in allowing easy endings. Sometimes it takes a government, as in the case of India, to step in and allow an alternative ending other than the ultimate one — death. Governments don't always get it right either. Some countries still require there to be a guilty party, insisting that something beyond the intangible dying embers of love must be responsible for the end of a marriage.

In the UK there are three main methods of divorce, according to the government's website. You can file a divorce petition, applying to the court for permission to divorce and showing reasons why you want the marriage to end. You can apply for a decree nisi and, assuming your spouse agrees to the petition, you'll be given a document proving this. Or you can apply for a decree absolute which legally ends your marriage. You have to wait at least 6 weeks after the date of the decree nisi before you can apply for a decree absolute, after which you are free to remarry.[15]

On the surface these provisions seem straightforward and reasonable, but dig a little deeper and the process reveals similar levels of denial that a marriage can end through normal human behaviour — simply by falling out of love.

To prove you have reasonable grounds for divorce you have to provide evidence that, as the government website points out, the court considers 'facts'. Imagine if we had to provide similar evidence at the beginning of a relationship? Providing 'facts' of love and commitment that show we can get married. Love being a subjective and emotional feeling, showing hard evidence would be difficult to come by.

There are five reasons which are considered reasonable grounds for divorce by the UK courts. None of them provides for an amicable, mature agreement between two adults that the relationship is dead. Instead they require evidence that one or another party has misbehaved, whether it's via adultery, unreasonable behaviour, desertion, the fact that you've lived apart for two years and have agreed to get divorced, or lived apart for five years without an agreement from one of you to divorce. What all of these so-called reasons overlook is that sometimes things just end naturally without earth shattering 'facts'.

An example of this culturally embedded denial of divorce can be seen in the UK case of Tini Owens, 66, and her husband Hugh Owens, 78. Having been married for 39 years, Tini Owens made 27 allegations of grounds for divorce from her husband. Saying he was *"insensitive in his manner and tone"*, that she was *"constantly mistrusted"*, he treated her like a child and she felt hopelessly unloved by him. The barrister defending Mr Owens clarified the UK legal position saying *"as the law stands, unhappiness, discontent, disillusionment are not facts which a petitioner can rely upon as facts which prove irretrievable breakdown."* Sir James Munby, hearing the case said *"It is not a ground for divorce if you find yourself in a wretchedly unhappy marriage - people may say it should be."*

However, Ayesha Vardag, a divorce lawyer interviewed by the BBC about the case argued *"judges should not compel people to stay married. This case highlights the absurdity of fault-based divorce. If a party is willing to go to the Court of Appeal to fight for a divorce, spending significant sums on the way, there is clearly no future for the marriage."* She added that it was *"beyond archaic"* that this situation should have to be proved to a judge.

Presenting *'facts'* to end something that emerged from emotive and subjective experiences shows significant bias in the balance of off-boarding and on-boarding. The way people fall in love is without logic or reason. Love starts and ends in mysterious, intangible ways. It's fickle and it shows very little logic. Often nobody is actually *'responsible'* for love.

The way we fall out of love is equally unpredictable, tedious and frequently without absolute blame. E-Harmony, which has 15 years' experience in online matchmaking has been pretty successful with it as they claim to see 438 members getting married every day. They believe there are ten major reasons why people fell out of love. [16]

1. They stopped communicating.
Conflict went unresolved, needs went unexpressed, and affirmations went unspoken. If good communication is key to building a healthy relationship, the lack of it can destroy one.

2. They took each other for granted.
It's easy to assume that love is unconditional and to get lazy with each other. When respect and kindness disappear, so can the love.

3. Expectations weren't met.
In the beginning it's easy to accommodate your partner's needs and wishes. Over time, however, people often default to 'just being themselves' and stop bending to the expectations of their partner, particularly when those expectations are not shared.

4. One of them discovered something new about his/her partner.
Betrayal can radically alter how someone feels. Discovering that your partner has hidden something from you, cheated on you, or behaved in a way that's inconsistent with who you thought they were, can do irreparable damage to a relationship.

5. Overwhelming jealousy took over.
Yes, it's reassuring to know that your partner wants you for him or herself. But when jealousy takes over, there's no room for trust.

6. The relationship wasn't built on a solid foundation.
If the relationship started poorly, moved too fast in the beginning, or was the product of an affair, it's likely not rooted enough to withstand time or overcome any real relationship obstacles.

7. Incompatibility.
As a couple gets to know one another better, and the initial fireworks die down, they may discover that their lifestyles, priorities, and values don't align.

8. Boredom or exhaustion.
The relationship has either lost its spark or become too much work for one or both people to handle.

9. A major life event changed things.
She's given birth and he no longer sees her as a lover, just as a mother. He got fired and suddenly retreats into depression and refuses her help. Instead of embracing life's adventures together, some couples crack under the pressures of hardships and break under the pressure of the unknown.

10. It wasn't love in the first place.
Lust can disguise itself as love. Once the honeymoon is over, however, it can also leave a relationship feeling empty and lacking. There's often a lot to be said for marrying someone who isn't just an object of lust but is also a friend[17]

None of these reasons seem uncommon, or unusual, but for UK courts, they would not warrant a divorce unless they fell into the narrow field of provable 'facts'. To be granted a UK divorce you need to show clear, facts-based blame. Your reasonable, mature, argument of the heart will not be tolerated - unlike the reasons to get legally married.

Grown up break ups

No-fault divorces are already available in many countries across the world. They aim to allow the dissolution of a marriage without evidence of wrongdoing by anyone. The first examples were legalised just after the Russian revolution in 1918, riding on the crest of a wave of utopian socialism, aimed at revolutionising every aspect of society, including long-held beliefs about divorce and gender.[18]

Many countries have adopted no-fault divorces to differing degrees since then. There is compelling evidence that it has helped people leave abusive relationships. It has even lowered the numbers of wives committing suicide by 8-16% according to research by economists Betsey Stevenson and Justin Wolfers, who also found a 30% decrease in domestic violence. No-fault divorces also allow couples getting divorced to create a more amicable post divorce atmosphere. This not only benefits their children, but also makes for a more positive

Shannon and Chris Neuman from Canada. via Facebook

atmosphere for dealing with what to do about income, homes, cars, pets and a whole host of other things we might acquire in the process of being married.[19]

Shannon and Chris Neuman from Canada were one of many couples to take this route and chose to tell an alternative story of a positive ending via selfies on Facebook. They took pictures outside the Alberta court house in celebration to tell everyone that you can divorce and enjoy a happy ending. In their words, *"We have respectfully, thoughtfully and honourably ended our marriage in a way that will allow us to go forward as parenting partners for our children, the perfect reason that this always WAS meant to be, so they will never have to choose."*[20]

Chapter 8

An intangible tale of services

The discipline of economics places products and services along a continuum. At one end of the scale are products – physical and tangible - at the other end are services. Services are identified by five broad characteristics. They are intangible. You can't store them. They are inseparable from the provider. They vary greatly in their delivery, often displaying inconsistencies. And lastly, the customer needs to be involved to experience them. These characteristics are called the 5'i's of services. - Intangibility, Inventory, Inseparability, Inconsistency, Involvement.[1]

As regards closure experiences, services present us with some interesting characteristics: the way we tell a story about services, the way we interpret them and (usually) the lack of physicality. Services come with historical baggage that characterises their relationship with closure, establishing an imbalance between the on-boarding and off-boarding of a service.

Short-term cultural commerce

What we experience as members of society, and as worshippers of a religion, we also experience in the narrative of commercial services that overlook or avoid the issue of endings. The services industry has experienced a boom in the last few decades as western economies move away from the traditional industrial model to a service model. According to the US Department of Commerce Economics and Statistics Administration, *"Between 1958 and 1992, total U.S. employment grew 100 percent (from 66 million to 121 million workers), while employment in service industries grew nearly 140 percent."* - growing from 48% in 1958 to 61% by 1992.[2] This trend continued throughout the '90s. According to the US Bureau of Labor Statistics 97% of jobs added to US payrolls between 1990 and 2002 were in the service sector.[3]

The financial services sector was once a respected and well thought of beacon of industry, but since the credit crisis of 2008 it has lost most of its allure. It has experienced some of the worst brand equity performance over the last few years, thanks to an endless stream of scandals: mis-selling, corruption and bad stewardship. This sector has hit an all time low and at the core of this low are dramatic changes in customer perception around trust. According to Deloitte, *"banking remains one of the least trusted institutions"*.[4]

Certainly the financial services industry has no shortage of stories around bad endings - PPI, sub-prime mortgages, mis-selling pensions - the list goes on. One terrifying example of this was published recently by the business website *Business Insider*, which showcased a small ad from

2005 advert for Simply Signature Loan.

real estate magazine from the summer of 2005 inviting customers to apply for a Simply Signature Loan. This, as the article mentions, *"shows just how loose lending practices became towards the end of the boom"*.[5]

The ad promises a speedy on-boarding experience. But as we now know, subprime lending like this proved hopelessly inappropriate. The company was simply out to sign up more customers. Usage of the loan exposed consumers to terrible long term problems, resulting in their ultimately losing their homes. It was a truly horrible conclusion and led to shameful off-boarding. The irony is that, as an industry, financial services sells many long term products to their customers, including pensions and mortgages. It would be fair to assume they'd be a whole lot better than average at delivering suitable endings to consumer experiences. In fact, the opposite is true.

Industry regulators end up with the messy job of sorting out the difference between promises made at the beginning of a consumer relationship and all the things that were simply not delivered by the service provider. In UK banking, it falls to the Financial Ombudsman amongst others to do this. The Ombudsman is tasked with settling agreements after normal communication has broken down between the consumer and the provider, both of whom have been completely unable to come to a resolution. In the last few years the Ombudsman's caseload has increased fourfold. In 2009 they considered 127,000 cases. In 2014 this leapt to 512,000. We could see this situation as an interesting bell-weather to the problem of unresolved endings.

The majority of the cases in the UK have recently been around Payment Protection Insurance - PPI. The UK Financial Conduct Authority calculates that £24.2 billion has been paid in compensation since 2011 in compensation. PPIs were very profitable to banks. The Regulator outlined in court how one of these products was sold alongside a mortgage *"costing the customer £20,838 over the loan term even though the maximum they could reclaim was £31,000"*[6] A investigatory panel on mis-selling and cross-selling established by the UK Government found that *"the prevailing culture within retail banks is focused on sales rather than on serving customers."* The consumer rights organisation Which?, who provided a great deal of evidence for the case from their own investigations, said that this *"...placed the achievement of sales targets over the long-term needs of the customer. Our recent investigation revealed that frontline staff are still under pressure to meet sales targets."*[7]

Cultural change doesn't happen overnight. The financial services industry is no different in this regard. So what happened to shift the industry to such scandalous short-termism, and its consequent detrimental effect on the customer lifecycle?

Measuring intangible relationships

Over the last few decades we have experienced big changes in our commercial banking relationships. Not so long ago the only banks were bricks and mortar establishments housing bank staff. Then things changed, leaving us with objective and systematic services which distanced off-boarding and closure experiences to maximise on-boarding of new customers for new products.

In the past bank managers were highly respected members of the community. This reflected the personal responsibility they had to local people. A recent paper from the University of Salford, by Pal Marathon Vik, describes them as the main *"providers of and gateways to financial services for households and businesses"*.[8] The bank branch had the autonomy necessary to provide cheque books, loans and mortgages. They also held responsibilities locally, which saw them sponsoring local teams or charities and sitting on boards of local businesses. Members of the community would value this meaningful relationship, where a genuine personal touch came with a personal responsibility to make sure the long term welfare of the customer was given priority. It wasn't about signing customers up to more services, about up-selling yet another loan. It was about understanding the wider picture and getting to grips with what the individual was trying to achieve in life. This was mostly intangible stuff and hard to capture on paper, but it is well understood by humans. It gave you a reason to stick with your bank. You knew the manager, they knew you.

The bank's' business narratives were changing fast and, on the back of the '80s' banking crash, banks introduced measurable techniques for checking people's capability to pay back loans. The introduction of credit scoring meant the removal of individual responsibility from banking staff and freed the manager from the responsibility of checking whether people could pay back loans or not. Vik describes the results as 'leading to a loss of the tacit knowledge about the local market that previously had been embodied in staff.'[9]

Where there was once a personal and possibly emotional relationship, a measurable and emotionless system now dominated at the turn of the millennium. This was fertile ground for the 2008 credit crunch. Numerous investigations revealed the short-term culture that drove us there: hard sales targets, limited options in the marketplace, and consumers being denied the mechanisms to move on to other providers.

A 2013 report by the UK's Independent Commission on Banking found that people changed their bank accounts on average every 26 years. It was believed that this was due to features like direct debits and standing orders, which took as long as 30 days to transfer to a competing bank (a full 29 days and 23+ hours longer than they take to set up in the first place).[10] This unwarranted delay prevented customers, many who lived from one month to the next, from switching accounts with regular payments to honour. This prevented customers from moving elsewhere and highlighted the denial that banking has long displayed with service endings. The report said this was responsible for trapping people in - often dysfunctional - banking relationships for far longer than they would otherwise stay. We were not trapped through loyalty to a close, trusting relationship with a bank manager, but because of the sheer difficulty involved in ending the service.

Paying off interest - or paying off debt

In the West financial services are familiar and accessible to almost everyone. This isn't the case across the globe, where access to simple loans can often be completely out of reach. Borrowing money has helped millions of people improve their lives and pull themselves out of poverty. The Grameen Bank, set up in Bangladesh by Muhammad Yunus, lends 'micro-credit' to the world's poorest people.

Despite lending money just like other banks, the Grameen Bank has a different philosophy around repaying loans. Their customers are encouraged to take pride in paying the loan back and they are celebrated when they do. In his book *Banker to the Poor*, Muhammad Yunus describes this philosophical change. *"The first time a borrower pays back her first instalment there is enormous excitement because she has proved to herself she can earn the money to pay it. Then the second instalment, then the third. It is an exciting experience for her. It is the excitement of discovering the worth of her own ability, and this excitement seizes her; it is palatable and contagious to anyone who meets her or talks to her. She discovers that she is more than what everybody said she was. She has something inside of her that she never knew she had. The Grameen loan is not simply cash, it becomes a kind of ticket to self discovery and self exploration. The borrower begins to explore her potential, to discover the creativity she has inside her."*[1]

In dramatic contrast, the Western financial culture praises us when we increase our loans, but it appears indifferent when we pay off our debt. Like the social narrative of marriage, we are often told it's good to start a commitment, and are congratulated when a marriage happens. Banks and loan companies congratulate us on a new loan, but react coldly or indifferently when we pay one off.

Unpaid until the last days

From one vulnerable group in the developing world being introduced to good debt and conclusive endings, to another vulnerable group in the Western world who grapple with the culture of interest only loans and under-considered endings. According to a recent report by the insurance company Prudential, one in four people planning to retire in the UK in 2017 will still owe money on their interest-only mortgage. This has risen from one in five in 2016. They found that on average, 2017's retirees owe £24,300 - an increase from the previous year by 29%, or £5,500.

Servicing a debt is common practice throughout our working life. Many of us pay off all sorts of debts. Our mortgages are possibly the largest and most important part of that. The common assumption is to do this before retiring so that you are then able to benefit from the reduced cost of living in your old age. Interest only mortgages are not a great use in this situation if they don't have some sort of pay out at the end to clear the debt. This traps borrowers in a loop of continuing loan repayments, without the conclusion of the mortgage being achieved.

A retirement income expert at Prudential, Vince Smith-Hughes commented on the research that *"For most people the move from work into retirement will see them having to cope with a drop in their income. So having to use precious retirement income to pay off debts could make life even more tricky for the newly retired."*

The research also found that 51 percent of retirees owed money on credit cards.

Credit cards

The difference between people investing for a well-thought-through capital gain and the emotional desire for something immediate has been exploited by the credit card industry. The way in which credit card loans are presented, and the structure for the repayment of debt varies hugely from the promotion of mortgages - thus creating an enormous imbalance in the customer lifecycle. Mortgage debt is presented as as having a definite conclusion. In contrast, credit cards and payday loans are presented emotionally, dealing with short term solutions and avoiding anything to do with a conclusive ending.

The average US household debt in the first quarter of 2016 was $132,086. Some debt, like a mortgage, is considered good debt (US average $171,775) because it builds up capital in your home. Other debt, such as payday-loans or credit cards (average US $15,310), are considered less good because all they do is provide freedom of access to money. [12]

In recent years, there has been a marked increase in the popularity of payday loans. These use very aggressive techniques to communicate their products to their target market. The advertisements promise easy access to money - but conceal the enormous interest charges - in some cases as high as 5000 percent APR.

Ofcom, the UK's communications regulator, found that advertising for pay-day loans had increased dramatically between 2009 and 2012 and *"accounted for 0.8% of all TV advertising seen by adult viewers. Very aggressive and persuasive techniques are used to promote these products to the target market. Who saw an average of 152 payday loan adverts on TV."*[13] Many of these advertisements targeted the unemployed, and people raising children: so inevitably they were seen by young children who happened to be watching TV when the adverts were on. Ofcom found that *"Children aged 4-15 saw 3 million payday loan TV adverts in 2008. This had grown to 466 million by 2011. By 2012, 596 million adverts were seen by 4-15 year olds, accounting for 0.7% of adverts seen by this age group."*

This seems pretty crazy. Are young children the incidental victims of the promotion to adults of payday loans? It certainly seems that way. This is darkly ironic considering how much parenting is based on reward or encouragement for finishing tasks - finish your dinner, complete your homework, tidy your room.

The comfort of credit

Credit card debt has been normalised, to the point that plenty of us have no real idea how much we owe. This erases a sense of conclusion in the process. According to research conducted by *NerdWallet*, a US based money advice website, *"Consumers vastly underestimate or under report how much debt they have. In fact, as of 2013, actual lender-reported credit card debt was 155% greater than borrower-reported balances."*[14] As the site points out, the reason for this disparity is partly because people are embarrassed by the levels of debt they have. Certainly, when questioned, 70% of Americans believe there is significant stigma around credit card debt, more so than other forms of debt. Thirty five percent said they would feel more embarrassment over credit card debt than a mortgage loan. Maybe some of the stigma is down to such debt being self indulgent. It has no functional or purposeful end. Compared with other types of debt that have less stigma, credit card debt has an endlessness about it, a circular feel, which suggests you have been investing in items of very little worth.[15] Mortgages, college fees, and medical debt arguably improve your life, and have a genuine goal attached to their ending. Tim Kasser of Knox College in Illinois, who has written about the subject in his book *The Price of Materialism*, argues that there is, *"an underlying attribution that the person with the credit-card debt has some sort of impulsivity, is unable to plan ahead or unable to follow through on responsibilities."*[16]

Game model

An old friend of mine, Lawrence Kitson, who has done more work with credit card, finance, and loyalty scheme companies than anyone should have to do in their life, provided some insights into the current climate for credit card customers. We started by talking from a personal point of view about some of the approaches he uses. As an admission, he framed himself as a pretty average user of credit cards.

"When I was 18 I agreed to visit the bank manager who wanted to offer me a 'financial review'. I ended up walking out with an overdraft and a loan. I've had residual debt on cards since I was 20, when I applied for an Egg card to buy some clothes for a holiday. They were funky cards, I liked their brand. They gave me way more than I needed. Over the course of the next 5 years the debt pile became slowly bigger.

I was the best kind of customer and a beacon of the cheap credit revolution in the UK of the late 90s and early 2000s. I took on debt, and didn't pay it off quickly. I was young and naive with no responsibilities and a limited income. So they made a fortune out of me. I was offered the best deals with the longest 0% available because they were betting that I wouldn't pay off before the end of the deal. So I was offered more cards and better rates. With no real understanding of what I was doing I cleared one card through a balance transfer and used the clear card for more purchases. The odd big purchase here and there, flights or festival tickets I couldn't afford, holiday car hire. That sort of thing. I just never paid it off before I bought something else

I've spent the last 10 years paying for this. I'm glad to say that pile is nearly gone. Nearly. The pile of debt that's left; it would be impossible to tell you exactly how I acquired it. I cannot say 'Ah that's that holiday paid off, good!' It's like a rotting pit of garbage decomposed beyond recognition, emitting fumes that waft into my life as interest."

As we continued our discussion he moved into his professional mode of designing services and talked about what a credit card, finance or loyalty company is looking for in a customer and how the business works. *"You get rewarded with points when you spend. To gain points you need to purchase something. The card company makes money at the checkout by charging the vendor when the card is processed. So getting you to make purchases is key. The items you might buy with the points you collect usually come as a result of a deal the credit card company has made with a supplier - an airline, restaurant chain or manufacturer for example.*

Credit card rewards, in my experience, are always opaque. The points are rarely directly equal to a currency. It would restrain the credit card company too much to do that. Instead offering 'rewards' helps them keep some fluidity in the game model they are using. This means that at one point a reward of a kitchen blender might cost 10000 points. Next year that blender might be worth 20000. In the shops it's always been $60. If they don't have the freedom to inflate or deflate their currency (points) they are restricted in what they offer, and what may encourage more usage of their payment and credit facilities."

I asked him about the inherent denial of endings in the debt model and how culturally we don't encourage endings. *"Credit card schemes don't want an ending, they are rewarding you to acquire debt, use their payment service by buying more items, and committing to their currency of points. I have never come across a scheme that rewards anyone paying their debt off, despite that being a good thing to do for many people. There is zero incentive to zero your debt. Debt is profit."*

Protection from the inevitable

One of the biggest financial service sectors is the insurance industry. Life insurance alone was worth $151.4 billion to the US economy in 2015, according to the Insurance Information Institute. That's a hike of 13.2% over the previous year.[17] Of course, very few of us want the conclusion of an insurance deal to happen. We naturally try to avoid car crashes, floods and illnesses. By insuring we are protecting ourselves against a potentially risky future, doing what we can to limit those risks. This is a good thing, of course, but when the services we've invested in have spent more time and effort getting you to sign up than they've spent considering the delivery and conclusion of the service, we have a problem.

The US insurance industry is currently under investigation for a lack of honesty with regard to payouts to family members of deceased customers. Investigations led by Florida State revealed that insurers had been using the social security *Death Master File*, a list of deceased people who were collecting social security retirement or disability income, to stop paying annuities when people died. But they failed to use the same file to activate a life insurance payout on a policy to other family members. [18]

Now 41 states in the US are investigating these practices. The Florida Office of Insurance Regulation said life insurance companies have agreed to pay out $7.4 billion. Five billion will go to beneficiaries who were failed by these life insurers. The remainder is going to States who are searching for people with unclaimed cases.[19] Perhaps surprisingly, plenty of states have established departments tasked with reconnecting policyholders with their missing money. The National Association of Unclaimed Property Administrators, for example, tracks down lost connections with bad service endings like the life insurance examples we've mentioned. A national database was established in November 1999, called missingmoney.com which was endorsed by the National Association of Unclaimed Property Administrators (NAUPA). In 2015 alone the organisation claimed it returned $3.2 Billion to rightful owners, from the $7.7 billion it has collected.[20] It has yet to find the remaining people who are out of pocket. Similar to the UK's Financial Ombudsman, the NAUPA wouldn't exist if we had better levels of consideration around the off-boarding of customers and took a healthier view of endings.

Reclaiming the perfect body

Gyms have some of the highest turnover of customers in any industry. According to *PT Direct*, a personal trainer website, as many as 30-50% of members. PT's analysis of the gym industry is based on research from the IHRSA (The International Health Racquet and Sports club Association) and The Fitness Industry Association (FIA), and it revealed some interesting aspects in regard to closure.[21]

PT Direct found that poor attendance by a new gym member would often get worse within the first month. People who attended more often in the early stages would improve to become steady gym goers, and if that behaviour made it to month 3, they would continue to attend regularly. Many gyms have responded to this high level of customer turnover with a form of entrapment, adopting aggressive rules in the contracts they offer. This involves unfriendly techniques like no get-out clauses, long lock in periods, automatic extensions, and no cooling off periods. *Which?* the UK consumer rights group, surveyed 543 British gym goers about their gym contracts in 2013. They found that 2 in 5 people had trouble cancelling their gym contracts. One in 4 people found they had a notice period of more than a month. And 1 in 5 found they couldn't leave the contract because they were tied in for a minimum term.[22]

Both the Federal Trade Commission in the US [23]and the Office of Fair Trading in the UK now provide guidelines to counter the culture of aggressive sales and unfair contracts with gyms. The Office of Fair Trading conducted an investigation in 2012 which pursued one gym operator called Ashbourne Management Services in the courts.[24] The company often relied on unfair contracts and debt collection practices, but the UK's High Court found that such practices were not fair on consumers, and presented an enforcement order to change the practices. The Office of Fair Trading is now establishing expectations for the entire industry in the UK.

What is surprising about these pieces of data is that together, they make a compelling case for ensuring quality closure experiences in the gym industry. The evidence points to a thoroughly disengaged consumer base who obviously like the 'feel-good-factor' of signing up and attempting to go to the gym. This actually represents an excellent marketing opportunity, but gyms are missing a trick. Instead, they are locking people in, damaging the relationship with their customers and undermining the brand. What is totally clear is that lots of people stop going to the gym. Overlooking that evidence is mad. Surely offering a really good closure experience would build a long term relationship beyond just one contract?

Where did I leave all that money?

One of the biggest and possibly most important investments you will make in life is focused on the end of it. Pensions take years to build up. As soon as we start work we're asked about our pension and how we intend to pay it. This long, drawn-out series of contributions that we make over most of our working lives is presented very differently from the exciting, often impulsive purchase of a holiday on a credit card.

The UK Pensions Commission believe a typical worker needs, *"income equivalent to about two-thirds of their final salary to maintain their lifestyle once they retire"*.[25] Given our expected lifespan - mine is going to be 86 according to the US Social Security website life expectancy tool - and the estimated retirement age of around 67,[26] I'll have to fund myself for 19 years till I shuffle off this mortal coil. Which, let's say on a 60k annual budget, means I need to save £1,140,000. It's eye-watering as well as highly unlikely. No wonder we're fed such good stories about pensions in our early careers.

The time between starting to pay into a pension and drawing on it in your later years is significant. The changes to a pension agreement over that time could be enormous. Companies change names, owners and locations. Governments change laws and pensionable ages, demographic time-bombs go off and your life might change beyond your wildest dreams... or nightmares.

In the last few decades working practices have changed dramatically, resulting in people having more, and a wider variety of jobs. Many countries have introduced separate pensions for every employer. The Department for Work and Pensions in the UK believe employees have on average 11 employers in their lifetime. That means 11 different pension pots with potentially different service providers, all needing administration for 50+ years.[27] It's a tough undertaking, not only for pensioners, but also for the providers. To keep in contact with people for more than half a century is quite an administrative undertaking.

Highlighting this, the UK charity Age Concern conducted a study. Their findings suggested that a quarter of these pension pots go missing, lost in the system over fifty or more years for no real reason other than the normal changes that take place over such a long stretch of time. Age Concern believes 47 per cent of these are 'simply lost in the mists of time', highlighting our dwindling interest in the end of a service.

By separating pensions into multiple pension pots from different providers, have we created far too many on-boarding experiences to balance well with just one ultimate ending? All of these pension pots are assumed to come together seamlessly, yet there is little evidence this is the case. We have encouraged more short-termism and clouded the delivery, overlooking the passage of time, the very thing we are trying to budget for in clear acknowledgement of our inevitable old age.

The undeliverable promise

One of the most epic endings we all fear comes with its own strange oxymoron: the nuclear deterrent. We are told of its clear benefits as a form of insurance. The UK's Trident program is a good example of a service provision that has never been delivered. And I'm guessing we all hope it never has to be.

The most recent estimated budget for a new UK nuclear deterrent is £205 Billion. [28] It's a Schrödinger's cat [29] sort of service. We are clearly aware of our inevitable end, we invest in being an active player in that end, but we think it's a solution to avoid that end. This story about protecting people and nations with deadly weapons has been around for more than six decades. It was first put forward by Bernard Brodie, the American military strategist, in 1959. He said, *"Thus far the chief purpose of our military establishment has been to win wars. From now on its chief purpose must be to avert them. It can have almost no other useful purpose."* He also pointed out that *"A nuclear weapon must be always at the ready, yet never used"*.[30]

It is interesting to acknowledge that we all exist under a terrifying threat of usage. Yet nuclear weapons are presented as a protection against an apocalyptic end, and it is a horribly compelling story for a service provider to tell its customers.

Too much fiction

We deal with services from the cradle to the grave. Their intangibility and other odd characteristics mean we tell stories as a way of defining those services. Questions obviously arise in all the examples quoted here, from the marriage story to the credit story. They all tell a great on-boarding tale, then fail to live up to the real ending.

In response to failed endings by companies, governments and religions, we give out fines, create laws, and change beliefs. Yet we still go on ignoring the validity of the end of these services as part of a better solution. I won't say that endings hold the answer to perfect services, but I do believe the less we consider endings, the more fiction we leave behind in our services. The solution is surely to acknowledge endings more. We should design, discuss and deliver better endings across all the services that touch our lives. At the moment the stories we are told about services do not reveal that they may have unconsidered, undelivered bad endings. Many are fairy tales that overlook the consumer's real experience of the ending.

'Turbulent', 2014. Unaltered Marine plastic found on the UK coast, 1994-2014.
Since 2007, artist Steve McPherson's primary source materials have been the discarded plastic objects that wash ashore as a pollutant on his local UK coast. While it is impossible to ignore the environmental concerns present in his work, Steve McPherson draws analogies with his practice akin to the role of archaeologist/collector/and paradoxical treasure hunter.

Chapter 9

Saying goodbye to your products

A passion for products

Our passion for possessions goes way beyond crude functionality. In fact, it goes back hundreds of thousands of years, far back into humanity's prehistoric past. By attaching value to objects through trade, exchange and debt, the earliest forms of ownership were established. The basic necessities of life from food to flints for weapons, to what was desirable - such as jewelry or clothes, and what was deemed to be property such as land or territory could all be classified according to varying statuses of ownership.

The sophistication of these items, their physicality, their meaning and their potential have matured over thousands of years to a point where our rituals of purchase, ownership and usage are incredibly rich in cultural meaning. Today, navigating this complex culture of consumer products fills a large part of our day. There is a lot to consider - the beauty in their physicality, the utility in their usage, the value in their materials, and the association of power and influence they represent.

Ends.

On-boarding today

Every month, according to *Mintel*, the world's biggest new products database, we see 33,000 different packaged goods including food and beverages, personal care and household products, clothing, tobacco, and pet food/pet care, introduced to consumers around the world.[1] And, on a larger scale, the International Organisation of Motor Vehicle Manufactures[2] reveals that 44,521,698 new cars were introduced to consumers around the world in 2015, in addition to the 907,051,000 currently on our roads.[3]

Relatively recently, an important addition to our purchases is the category of consumer electronics: things like radios, TV sets, MP3 players, stereo systems and DVD players, desktop computers, laptops, tablets and smartphones. Before 1990 this category accounted for a fraction of consumer goods, it now accounts for 210 billion US dollars' worth of new goods every year.[4]

Regardless of their category, every one of these products has had an enormous amount of effort applied to its creation in the shape of meetings, designs, models, proposals, pricing plans, prototypes, testing, assembling, shipping, marketing and advertising. That process employs millions of people across the world at any one time, all dedicated to getting an emotional reaction and financial commitment from the consumer.

The ownership experience

Ownership is an important experience for the consumer, providing them with the empowering feeling of freedom, control, and the ability to do with the product what they desire. With regard to closure experiences and endings, it is also important to consider how ownership is transferred or ended, since it influences the off-boarding of the consumer experience.

Broadly speaking there are three types of ownership - common property, collective property and private property. Common property refers to items that are available for use by all members of society. It might be a tract of common land that people use to graze cattle, gather firewood or food. It could be country parks, recreation grounds or nature reserves, places that also fall under the 'common' definition.

Collective property is owned by a group of people or a community, and they determine how this kind of item of property is to be used. Decisions about this might take place through some sort of collective discussion, for example elders in tribes and communist groups in China.

Private property comes under the sole interests of an individual, group, family or firm, to do with as they wish. Despite this, private property still requires societal norms. If property is stolen, for example, the owner will expect society to step in, in the form of the police, to take responsibility for retrieving the item and upholding the law.[5]

Defining the moment of waste

A great deal of consideration, thought and legislation has been devoted to the relationship between the consumer and the moment of purchase, in stark contrast to the consumer relationship at the off-boarding stage, the moment of waste-creation.

Apart from the disposal of poisonous or dangerous waste, the definition of ownership moves quickly moves from individual responsibility to a societal concern. The abrupt transition in the object's status from 'my product' to 'society's problem' arguably frees the consumer from the wider consequences of consumption. Because it takes but a moment to off-board ownership of an item, the opportunity for reflection and responsibility is removed.

The Organisation for Economic Co-operation and Development (OECD) defines waste as "...*substances or objects which (i) are disposed of or are being recovered; or (ii) are intended to be disposed of or recovered; or (iii) are required, by the provisions of national law, to be disposed of or recovered.*"[6]

This is a cold and functional definition, and it isn't uncommon.

Plenty of organisations use a similar approach, completely ignoring the emotional journey that consumers have made when purchasing and using a product. This is also an approach that's questioned by Seiko J. Pohjola, and Eva Pongraca of the Chemical Processing Engineering Laboratory, at the University of Oulu in Finland, who encourage a more emotionally accurate version and a firm attachment to the consumer's role, "...*waste is something that the holder has disposed of/discarded or is going to dispose of/discard. While 'dispose' and 'discard' both mean 'abandonment', 'disposal' indicates perhaps that the waste is to be put into a suitable place, whereas 'discard' comes over more as something being useless or undesirable, which needs to be tossed aside..*"[7]

Pohjola and Pongraca go on to propose a redefinition of waste which will create closer ties to individuals; something they visualise as falling into four distinct categories, four classes of waste.

Class 1: Non-wanted things whose creation was not intended, or not avoided, and which have no purpose

Class 2: Things that were given a finite purpose, thus destined to become useless after the purpose was fulfilled

Class 3: Things with a well-defined purpose whose performance is no longer acceptable

Class 4: Things with a well-defined purpose and acceptable performance, whose users failed to use them for the intended purpose
[8]

This approach to connecting waste and the consumer is also championed by Ilona Cheyne, Professor of Law at Oxford Brookes University, and Michael Purdue, University of Newcastle upon Tyne, who provide an insightful argument around the moment of disposal, "*The essence of the legal definitions is that the owner does not want it; thus waste exists only where it is not wanted.*"

The big landfill issue

Once a product is no longer deemed to be useful, the consumer very quickly perceives it to be waste. In an instant we distance our consumer selves from responsibility for the product, as we drop the item in the trash. This disposing process has totally different characteristics from the early part of the customer life cycle. In contrast it is emotionless and cold, avoids any opportunity for reflection, and gives up on the responsibility of ownership.

Solid waste is a massive global problem with enormous consequences for the environment, not only in the physical location of waste but in the wider knock-on consequences of climate change, water pollution, and loss of resources. The magazine *Nature* explains the scope of the issue. *"As urbanization increases, global solid-waste generation is accelerating. In 1900, the world had 220 million urban residents (13% of the population). They produced fewer than 300,000 tonnes of rubbish (such as broken household items, ash, food waste and packaging) per day. By 2000, the 2.9 billion people living in cities (49% of the world's population) were creating more than 3 million tonnes of solid waste per day. By 2025 it will be twice that —enough to fill a line of rubbish trucks 5,000 kilometres long every day."*[9]

Waste and landfill challenged earlier generations as they experimented with various initiatives to deal with it over the centuries. Early attempts saw us re-use items, pass hand-me-downs to other members of our family, and mend broken products. But these have fallen away in favour of recycling and disposal to landfill. Both are are quicker routes for the consumer to off-board from a product, and both quickly make way for the purchase of new items.

In their book *Cradle to Cradle*, Michael Braungart and William McDonough describe the status quo of manufacturing and the consumer experience as having an enormous disparity between an item's assembly and usage, and alternative ways of disposing in landfill or recycling. *"Cradle-to-grave designs dominate our modern manufacturing. According to some accounts more than 90% of materials extracted to make durable goods in the United States become waste almost immediately. Sometimes the product itself scarcely lasts longer. It is often cheaper to buy a new version of even the most expensive appliance than to track down someone to repair the original item"*[10]

Alternatives to the dump

Some countries are attempting to widen our approach to off-boarding and bring back historic alternatives to disposal. Sweden is aiming to encourage mending of the nation's worn out or broken shoes, clothes and bicycles by slashing the VAT on repairs from 25% to 12%.[11] The proposal will also let people reclaim the cost of repairs made to fridges, dishwashers and washing machines via their income tax. Sweden's Minister for Financial Markets, Per Bolund, is hoping for a sea-change in the behaviour of the country's citizens, believing, "*This could substantially lower the cost and so make it more rational economic behaviour to repair your goods.*"

Alternatively, reclaiming the materials that make up disposed-of products before they become landfill is a helpful method of reducing waste. Take-up of this approach varies between countries. According to *Nature* magazine 'Japan issues about one-third less rubbish per person than the United States, despite having roughly the same gross domestic product (GDP) per capita. This is because of higher-density living, higher prices for a larger share of imports and cultural norms.' These cultural norms can also be seen in Japan's approach to recycling. The Japanese town of Kamikatsu, for instance, separates its waste into thirty four different categories, an enormous undertaking for any resident and quite a contrast to the three or four categories consumers are tasked with in other countries.[12]

In the 1990s Kamikatsu dealt with the issue of daily waste by burning it in open fires. Finding this unwieldy and dirty, they decided to invest in an incinerator, but sadly found the model they picked was outdated as soon as it became operational. Confronted with such disappointment, the town felt they had to deal with waste differently from simply sending it up in flames, so they embarked on the ambitious plan of 100% recycling by getting their residents to sort their waste into 34 different categories. To date Kamikatsu's residents recycle about 80% of their rubbish, with the aim of achieving 100% recycle rates within the next 5 years.[13]

The people of the town are confronted with the same issues many of us have with recycling, which is interpreting the materials we discard. To resolve the issue, there's an expert employee at the processing centre who helps them decipher and classify the materials used in the objects they are throwing away. As consumers they are becoming more aware of materials and the ease or complexity they might have in the recycling process.

The small amount of discussion it take to involve the consumer reflects the lack of consideration the issue has been given as a 'user experience' in the first place, earlier in the process. Most of the focus has gone into the beginning of the customer lifecycle, the on-boarding stage. The end of that life-cycle has been left for individual interpretation, quite literally in Kamikatsu's case. The closure experience for the objects we consume is now deciding what the object is actually made of, so as to determine what to do with it.

Bio belief

A route that promises less burden to the consumer but which would have a similar environmental benefit is the idea of the circular economy. The Ellen Macarthur Foundation, a big supporter of the approach, says, *"a circular economy is one that is restorative and regenerative by design, and which aims to keep products, components and materials at their highest utility and value at all times, distinguishing between technical and biological cycles."*[14]

Although many people have explored an approach like this, it has rarely been as well framed as it was by the chemist Michael Braungart and architect William McDonough in *Cradle to Cradle*, which contains a host of compelling stories about businesses adapting to a circular economy approach. By doing so these businesses not only made their products far more environmentally stable, but also increased efficiency in manufacturing and boosted their finances. Sadly in this book there are very few examples of what this means as a consumer experience.

A circular economy approach happens to a large degree behind the scenes for the consumer.

The difference between purchasing a chair with a good circular economy process against a chair manufactured in a traditional way is difficult for the consumer to perceive. Removing the consumer even further from the discussion about waste, and consequently about their responsibility surrounding consumption, is a lost opportunity. Yes, a sustainable approach is badly needed. On the other hand it changes very little in the consumer experience status quo.

Sadly Braungart and McDonoughs' subsequent book, *The Upcycle*, moves even further away from the consumer experience. It even champions leaving the consumer experience as it is - in denial about the opportunity for reflection. This is characterised by former president Bill Clinton's foreword to the book. *"The optimist says the glass is half full and the pessimists says it's half empty. Bill and Michael say it's always totally full-of water and air-and they are constantly working to share that full glass with more people, to make it even bigger, and to celebrate the abundance of the things that enable us to thrive."*[15]

Adapting the current manufacturing processes to make them less damaging to the environment is to be celebrated, but this needs to work in partnership with growing consumer awareness, not simply indulging our denial and supporting the notion that our actions have no consequences.

Old mobiles in drawers

E-waste is a huge problem, and one which doesn't receive the attention it should. It is also a relatively recent problem, having really only been around on a massive scale since the 1990s. Since then we have taken on new tech with a passion, and the sheer quantity of our purchases is increasing fast.

This complex issue demands a variety of responses to achieve reduction. One surprising route, given the rapid turnover of electronic products, is mending and upgrading them so the need to replace is reduced. One group leading the way in this is the Restart Project. The project's founders, Janet Gunter and Ugo Vallauri, are attempting to change our perspectives on the death of electronic products, encouraging people to breathe new life into broken electronics through free community events, which provide support, tools and parts. Their aim is to intervene before people dispose of their electronic devices, thus reducing the burden on the world's fastest growing type of waste.

The US population alone disposed of 384,000,000 devices during 2010, but only 19% were re-cycled. According to the United Nations University's Institute for The Advanced Study of Sustainability, 42 million tons of e-waste were generated globally in 2014, of which about six million tons was ICT related. These numbers are growing aggressively and developing countries are expanding the market. China, for instance, has become the 6th biggest ICT market in 2017, growing from just 1% computer penetration in 1998.[16]

Another worrying aspect is that e-waste product lifespans are getting shorter. When consumers would use their tube-based TVs for over a decade, the replacement cycle is now believed to be closer to 5-7 years.[17] Older mobile phones would last around 4.7 years[18] but smartphones seem to live for just 21 months.[19]

Restart Project

For many consumers, the off-boarding of a electronic product is a complicated process. Recycling isn't easy. Dis-assembly is impossible for amateurs thanks to the sheer complexity of the products. Curbside waste pick-ups from councils rarely take them, so people have to make a special trip to a recycling site that takes e-waste. Most consumers know they shouldn't put phones and other electronic gadgetry in the regular waste bins, but they don't know what alternative action to take. The sad thing is that many of these devices could live longer with some basic knowledge, a few tools, the right attitude and a little encouragement.

Talking to Janet and Ugo about the Restart Project provides fascinating insights into the consumer lifecycle. Although Restart events are set up to fix devices, they do so much more than that. It's almost like a support group for cheated consumers, whose anecdotes are astounding. Janet and Ugo hear hundreds of stories about terrible customer service, poor assembly of parts and mis-information from manufacturers. If you work in the tech sector, you could learn a great deal about consumers by attending one of their events.

Breaking the warranty seal is often the first thing that Janet and Ugo encourage people to do. Although the warranty label stuck inside electronic devices actively discourages tampering, it systematically discourages the re-birth of these products with its foreboding message of fear. To mend the device, they have to break the agreement with the manufacturer, in essence bringing a proper closure to the relationship so the device can live on freely as a hack.

As Janet says, "*People feel un-empowered about dealing with technology. Some feel it's almost a conspiracy. And they are looking for permission to mend their device. They have a perception of us as experts, who can offer impartial and knowledgeable advice. Once we teach them how to take a product apart, it changes their knowledge of what is a good product.*"

Beyond physically mending gadgets, Janet and Ugo also end up offering plenty of consumer rights advice and simple encouragement on a very different level. Sometimes people bring in devices that haven't been used for ages, the consumer having moved on to another device and neglected to deal with the old, broken one, left in a drawer until they can resolve the issue. In effect the consumer has closed the relationship with that product and moved on to another.

Fixing tech is also good, Janet explains, because, *"they get handed-on or sold. But the important thing is that they displace another purchase - It seems common with food blenders."* On the other hand there are some terminal issues that just isn't worth attempting, according to Janet. *"If the motor on a vacuum cleaner is gone, basically, it's dead, and the discussion then becomes one about where and how to recycle it."*

Changing laws

Consumers who own electronics are not the only group to feel frustrated at being denied access to their broken products. The US auto industry has had the same issue for over 10 years with independent repairers and suppliers lobbying the government with a *"right to repair"* bill, that would change the law and allow them access to diagnostic tools and repair data. Until the state of Massachusetts passed a *'Right to repair'* bill in 2013, such access was available only to car companies and their franchises.

A year later, two of the biggest trade groups representing the Alliance of Automobile Manufacturers and the Association of Global Automakers agreed a deal to make the law nationwide across the US. This will provide universal access to diagnostic codes and repair data in a common format by 2018.

Another group which feels excluded from making their own repairs are US farmers. Many of them carried out their own repairs in previous decades, but more recently have been restricted by warranties, access to diagnostic software and repair data. Many own machinery worth hundreds of thousands of dollars. When these go wrong they have to call the manufacturer or a franchisee, for a repair as they have unique access to the tools required. This results in long waiting times for for getting repairs done, plus eye-watering bills. This is something that independently-minded farmers find extremely frustrating.

A lobbying group have put together a proposal for a *"Fair Repair"* bill. Nebraska is the first of eight states - Minnesota, New York, Massachusetts, Illinois, Wyoming, Tennessee and Kansas, to hear the legislation on March the 9th 2017. This would allow access to the diagnostic tools and machinery in the same way as the Right to repair' bill did.

One significant difference is that this bill will impact well beyond the farming world. It is likely to awaken the dragon of the technology industry, which has, until now, restricted the repair of phones and other electrical goods to themselves. This has limited opportunities for the owner of electronic goods to get them mended in the free market, or even repair them themselves.

The technology industry has, of course, a lot more power and influence than the big agricultural companies like John Deer. Many of the backers of the bill are concerned that the influence and clout of the tech giants will overwhelm their voice. Certainly the sponsor of the bill, State senator, Lydia Brasch, is under no illusion about the opposition, saying she had *"never experienced lobbying like it"* when speaking to Apple about the issue.

Saying goodbye

When we buy a product, all sorts of marketing messages, point of sale at checkouts and advertising campaigns play their part with a low-level introductory greeting. Although the word 'Hello' might not actually be used, many starting experiences mimic the warm emotional etiquette that we use between ourselves every day. In Apple's case it was quite literally a 'hello', and it worked to great effect when introducing the Macintosh to the dry and emotionally cold PC market in 1984.

In contrast, emotional 'goodbyes' are rarely used at the end of a product relationship. Business tends to be repelled by the end of the product relationship, choosing to drag the customer back to a new 'hello', ignoring the 'goodbye' stage entirely.

Macintosh 128K, Apple Macintosh, vintagemacmuseum.com

This is probably the wrong approach, especially if consumer engagement at the off-boarding point or even a positive action is to be encouraged. We know how powerful emotion can be at the beginning, but we rarely use it at the end.

One individual countering this trend is Marie Kondo, the self-proclaimed declutterer. She has witnessed every variation of the imbalance between gathering items and disposing of them. When it comes to saying 'goodbye' to products, few people have more knowledge. She has developed the Marie Kondo Technique of Tidying on the basis of years of obsessing about tidiness and consulting on the cluttered homes of clients. Her book *The Life Changing Magic of Tidying* has sold over a million copies and reveals the heart of the technique as the emotional engagement with items that a person has hoarded over years. Saying 'goodbye' meaningfully is key to the technique. [20]

Hoarders are particular victims of an imbalanced customer lifecycle. Their emotional engagement at the beginning of the customer lifecycle is guided quickly to a purchase, and rewarded through ownership and usage. Thereafter guidance is lacking and products tend to linger, paralysing cupboards and drawers and leading to homes so full of stuff that they are no longer homes, more like landfill sites.

According to Regina Lark, another professional organiser, the average US home contains around 300,000 things, from ironing boards to paper clips. Thanks to all the stuff that we find so hard to get rid of, we often need to look for additional space. Small surprise, then, that the New York Times found the fastest-growing element of the US real estate sector is off-site storage. Apparently 1 in 10 Americans use it, highlighted by a flurry of TV programmes about entrepreneurial types buying and re-selling the contents of abandoned storage units.[21]

Kondo believes that storage isn't the answer, but part of the problem. *"Putting things away creates the illusion that the clutter problem has been solved, but sooner or later all the storage units for the room once again overflow with things, and some new storage becomes necessary, creating a negative spiral. This is why tidy must start with discarding."*

Discarding might seem like a bridge too far for a consumption-happy consumer, but for some hoarders clutter is a serious issue. It can cripple their home life, creating a difficult, almost antisocial existence. Trapped in a house full of things that they can't get rid of for one reason or another, these people are of course extremes on the boundaries of behaviour, but many of us are suffering the same kind of thing at some level. If you open a cupboard at home and everything falls out because it's crammed so full, you might be one of them.

Marie Kondo can relate well to these psychological issues. She has experienced years of personal obsessing, and some very difficult emotional situations with items that she owned.

A profound breakthrough came after a big bout of discarding unwanted items left her particularly depressed. While lying on the floor in her room, exhausted from throwing old products in bin bags, it occurred to her that she hadn't left space for the things she wanted to keep. The feeling of throwing things away was emotionally barren when it should have felt uplifting and liberating.

Marie hadn't stopped to consider which items she cared for. And with that, she realised she needed a new approach to decluttering, a way to assess each item individually and emotionally. She came to the conclusion that 'the best way to choose what I want to keep and what to throw away is to take each item in one's hand and ask *"Does this spark joy?"* If it does, keep it. If not, throw it out.'[22]

Many years of working with hoarders and championing the Marie Kondo Technique have left her with a surprisingly emotional relationship with products. She certainly values what she has more than many consumers, showing a rare emotional connection that many of us would find childlike and bizarre. Even an everyday situation like arriving home is full of emotional reflection, revealing her unique approach to the value of objects and the lengths she goes to appreciate them. *"First I unlock the door and announce to my house 'I'm home!' Picking up the pair of shoes I wore yesterday and left out in the hall I say, 'Thank you very much for your hard work,' and put them away in the shoe cupboard. Then I take off the shoes I wore today and place them neatly in the hall. I put my jacket and dress on a hanger and say 'Good job!' I put my empty handbag in a bag and put it on top shelf of the wardrobe saying 'You did well have a good rest'."*[23]

Although Kondo's behaviour might seem strange to us, maybe we should think again about why it doesn't sit comfortably. It's not unusual to use emotion at the beginning of the customer lifecycle, so why should we find using emotion so strange when we are using products or disposing of them at the off-boarding stage? Some of Marie's clients talk evangelically about her technique and its positive impact on their lives. It can mean freeing them to achieve things that they previously only dreamed of, maybe starting businesses, leaving a loveless marriage, changing their outlook on life. What can we can take from this? It appears that the power of personal decluttering isn't just about getting a lovely, uncluttered cupboard, but about improving your life fundamentally. It might be more important to appreciate the value of a product with a meaningful farewell than to begin another product relationship, acknowledging the importance of saying 'goodbye' to old items over saying 'hello' to more items. Or as one of her clients put it, *"letting go is even more important than adding."*[24]

Ends.

The Marie Kondo technique

The first problem is one of visibility. We often can't see all the stuff we own. It's tucked away in drawers and cupboards, on shelves or stored away in the garage or loft. Some of it might even be at a storage facility.

Kondo gets her clients to gather all the items of a single category together. This highlights how much time and effort we spend buying yet another item in the exact same category, and just pushing the old one to the back of the cupboard. We do this for years, accumulating enormous quantities of items and never assessing them at the same time, as a single entity. Seeing it all lumped together lets us judge how much we own more accurately. For many it is revealing. For others it's actually embarrassing. But once all of these items are displayed together, the individual can start to assess what really sparks joy: picking up each item, feeling it, and seeing if it brings that unmistakeable emotional jolt. Kondo often has to challenge her clients initially, because they're so wrapped up in the 'I might need it in the future' mindset. They often find it hard to tune in to the emotional side of their feelings around objects. But once they get it, they quickly start to make progress in de-cluttering.

The next stage of the process, once people have worked out what doesn't bring them joy, is to thank the items they'll be disposing of. Kondo points out that we can even thank items that have simply taught us something. *"Why did you buy that particular outfit? If you bought it because you thought it looked cool in the shop, then it has fulfilled the function of giving you a thrill when you bought it. Then why did you never wear it? Was it because you realised that it didn't suit you when you tried it on at home? If so, and if you no longer buy clothes of the same style or colour, then it has fulfilled another important function - it has taught you what doesn't suit you. In fact that particular article of clothing has already completed its role in your life, and you are free to say 'Thankyou for giving me joy when I bought you' or 'Thankyou for teaching me what doesn't suit me' and let it go."*[25]

Thanking various gods

Is Kondo onto anything new? Is saying thank you to inanimate objects a meaningful step along the way to improving our approach to consumption and creating better closing experiences at the off-boarding phase in the customer lifecycle?

Of course, expressing versions of thankfulness, gratitude and appreciation form a big part of all the world's religions. We could say that they are almost central to a religious belief system. Many of us are familiar with thanking various Gods for being healthy and happy, showing appreciation for others or even thanking those Gods for the gifts that we have. But Kondo is thanking the item itself, not thanking a deity for the item.

In the last hundred years poverty has been reduced, child mortality has decreased, and life spans increased. Yet many religions still involve being thankful for issues, and they're not always necessarily modern ones. Choosing to give thanks for the presence of a meal can seem outdated in a world of Western obesity.

Religious appreciation - giving thanks - has been lost due thanks to the erosion of a believable context. That context was once starvation. Now the context is abundance, and we have a consumption crisis on our hands that has resulted in climate change. We should probably be saying grace when we avoid taking a carbon-impactful flight, and use Skype instead.

Festivals like Christmas, which traditionally gave thanks, have changed into an exercise in mass consumption, reflected by the businesses that are made or broken at Christmas. The New Year financial news stories often feature the heavily anticipated results of companies' Christmas sales. Organisations operating globally have several festivals to make profits from, for example Thanksgiving, Chinese New Year and Diwali. After a short four month period between Diwali in October and Chinese New Year in January, plenty of companies can tell whether or not they've been successful for the whole year. The creation of fictional festivals like Black Friday, which openly celebrate consumption, has only increased the intensity of the constant pressure to consume.

We shouldn't forget however, the other aspects of Kondo's approach. Take the creation of animism in the product, when she speaks to the things she owns, and disposes of them as if they had feelings. Many religions believe people have a soul or spirit, but fewer believe that objects possess some sort of Animism, which was a popular belief in primitive cultures, dying out with the introduction of more organised religions. Buddhism and Shinto, however, do still respect components of animism in their religions. They demonstrate that respect with funerals for a wide range of animals and objects, everything from from pets and dolls to chopsticks and mobile phones. The aim of these ceremonies is to give thanks to the spirit of the item, in some cases to protect the owner of the object from upset or bad luck, in others to show thanks for a job well done. Hari Kuyō the 'funeral for needles' ceremony examples this. The needles are placed on a piece of material, often tofu, in the belief that the objects should have a comfortable final task of piercing something soft, after a lifetime of piercing hard objects. According to Angelika Kretschmer, who has observed some of these rituals *"the disposal of old objects plays such a prominent part in these rites, we can say tentatively that, for the individual, the motivation might be an attachment to a particular object and an unwillingness to simply throw it away out of respect."*[26]

Maybe a more famous festival is the 'memorial for dolls', honoured annually shrines across Japan. One of the most famous shrines is the Awashima Kada Shrine in Wakayama, which receives an astonishing 300,000 dolls every year.[27] This level of respect is characterised nicely in a interview in the Japanese Times of an elderly recent widow, who brought two dolls along to a 'Memorial for Dolls' service. The couple had received the dolls as a wedding present over forty years ago. She believed just throwing them away would bring 'divine punishment', and 'wanted to have them prayed for'. Having a product for four decades should generate a bit more respect in our hearts, giving it an emotional and meaningful farewell.[28]

A warm send off

The imbalance of the customer life cycle between an overwhelming on-boarding experience and muted, sterilised off-boarding experience can be adjusted through the use of emotion and appreciation. Saying *"Goodbye"* at the end of a product's life shouldn't seem strange. We put emotion and meaning into our *"Hello's"*, yet sterilise our *"Goodbyes"*.

A good closure experience should have real meaning. The purchase had meaning when we said *"Hello"* to the product we bought. Adding meaning will bring appreciation. Bringing appreciation to our product ownership will mature the discussion about consumption and might even bring alternatives to the sometimes frightening consequences of consumption we face today.

Monument Valley is an indie puzzle game
developed and published by Ustwo Games

Chapter 10

Closure in digital

Some time in the '90s I signed up to my first digital service. It was an internet service provider (ISP). There weren't many around then, maybe five or six providers throughout the UK. I remember the on-boarding being so crude that I had to make several phone calls and complete a contract by fax.

Twenty years later I am still signing up to digital services, but now it's faster, smoother and mainly via my phone. I hardly remember doing most of them, it's so effortless. In the intervening period I must have potentially signed up to hundreds of different services, many of which have since fallen by the wayside, been acquired by bigger companies, changed names or gone bankrupt.

In some cases I was the one that made the decision to change and stop using the service. Sometimes I moved from one platform to another, in some cases I may have upgraded in the process, or moved to another provider. But quite frankly I don't know what has happened to most of these digital relationships. I know that once I was in them, and now I don't use them. They've been lost without a proper goodbye, ended without a conclusion, and I've left my personal stuff, content and details behind.

Ends.

As a consumer I don't care. I am too excited about the next digital experience. But as a person who wants to retain a life-long reputation, I should probably think more about the consequences of leaving a legacy of lingering assets online, a trail of unfinished business.

The end of many digital services happens without a coherent conclusion. Ending one relationship is often the consequence of starting another. Some of these services coerce consumers into lifelong participation through photos, comments, video and other content, all available to a global public as a result of endless encouragement to share, sign-up, provide details and hand over data. Much of the information provided by the consumer might expose them personally in the wrong situation in the years to come, thanks to unwanted evidence online that is completely out of their control, lingering long after they have left the service they signed up to.

There is no doubt that the benefits of a digital world have been enormous. Our lives have all been improved in one way or another, directly or indirectly, thanks to its remarkable capabilities. But as with previous industries that enjoyed dramatic periods of growth, blinded by the continual drive forward, we overlook important aspects. Digital is no different. It has driven forward, changing the fundamentals of society, old behaviours, laws, etiquette, education and even our methods of remembering and capturing history. All these need to be redefined for this new construct. We're all part of this brave new digital culture, a young culture fumbling around trying to pin down a definition and a vocabulary of its values and principles. We're all engaged in trying to work out what is best for society.

Sadly, and not unexpectedly, we have brought assumptions from previous industries with us and tried to map them onto the new landscape. The bias of the customer lifecycle is one such assumption. It favours on-boarding, assuming that only the short term benefits matter and leaving us overlooking the long term consequences. The question of how we conclude many of these digital services in a neat and tidy way that protects our identity and acknowledges privacy is often left lingering, leaving a pollution of digital assets behind us.

Closure experiences and the harmony between on-boarding and off-boarding are pertinent subjects in this regard. Left unchecked, these might establish themselves as the norms in our new digital society.

Bomb proof storage and infinite memory

The emergence in the mid 20th century of the atom bomb and the increasing reliance on computers exposed America to the risk of losing crucial data. It was feared that a critical nuclear attack by a foe would wipe out the nation's entire computer system - and remove an essential pillar of America's capability. So a more robust and less localised protection for computer data was required.

In 1973 the US Defence Advanced Research Projects Agency, DARPA, commissioned a project to develop protocols which would allow computers to communicate across multiple linked packet networks. It was called the 'Internetting project' and what emerged was the 'internet'. This new space was made up of a growing network of interlinked remote computers that mirrored themselves continuously, and therefore never lost data. If one computer was removed, others would take its place.[1]

Over the last forty years we have been adding to, and building on this system. The original brief was to build a distributed data storage system that would be resilient to nuclear attack, a system that would never forget or lose the things we placed in its keeping. Both the data we put on it and the system itself are bomb proof.

The internet is arguably one of humanity's most successful creations. Its growth has been impressive. Within a 40 year period we have managed to get nearly half the world's 7.3 billion population onto it, converting 3.4 billion individuals into active internet users, a penetration of 46%. Two point three billion of these are also active social media users, a penetration of 31%, and the majority of these

(1.9 billion) are accessing the web via mobile technologies, now at a staggering 3.7 billion unique users. [2]

Those born in the last two decades are naturally digital natives. They experience this connected world as borderless, empowering and dynamic, an amazing place to be with others, interact and present yourself. The Pew Research Centre, says that 1 in 4 teens are online almost constantly and 56% are online several times a day, mostly thanks to easier access through smartphones and the apps that accompany them. [3]

A large proportion of the content we create every day as users are photos, documenting the details of our lives, our dinners, our cats, our abs, our loved ones. One point eight billion photos were uploaded online every day in 2014, three times the previous year's figures according to *Mary Meeker's Annual Internet Trends*. Given the typical annual increase, that number is probably more like 6 billion as I write, and by the time you read this book it will have gone up again. [4]

More and more news is being consumed and shared with friends online. According to the Pew Research Centre, "*62% of U.S. adults overall now get news on social media sites*". And watching videos and films is also increasing, with over 300 hours of videos uploaded every minute to YouTube. We already watch 4,950,000,000 videos every single day, and the numbers are increasing by the the month. [5]

Cultural short-termism

What are the basic activities involved in social media? These could include videoing some remarkable incident, snapping a photo, reading an up-to-date news story, maybe even devising a witty chunk of prose just 140 characters long....and then sharing them with our friends and followers. Such activities sit at the heart of modern communication. Most companies provide a way to share these actions at every point in the interface: a photo app or gallery on your phone, a video editing app on your iPad, or an article on a news site. Wherever we go online, we're encouraged to share. But we've lost the precious ability to create endings. Where does my content end up, who knows my identity, who is watching me, and how can I stop that video being shared? The answer to these questions is often *"Who knows?"*.

Our lives, attitudes and behaviour change as we get older. One day we will have to deal with the growing chasm between a human life-span and a digital life-time. A flippant remark about a friend's school clothes might linger in our profile for decades. A picture taken in a drunken moment with a boyfriend or girlfriend might cause endless regret when strangers contact you about it.

We aren't going to stop these behaviours. We won't stop engaging with new ways to communicate, feel closer, and share the things we feel are important with a wider audience. But we can improve the levels of control, empowerment and safety we feel at the end of an experience, making improvements in the ways we remove inappropriate information or keep it more private. While we have simplified the process of starting these relationships efficiently, we have almost completely overlooked how to stop them. The digital industry is in denial about endings.

Anti-amnesia

Evolution has enabled humans to develop a well-functioning memory that normally - in time - fades. Digital has provided a foolproof, perfect, globally retrievable memory that won't forget. Ever. In his book *Delete*, Professor Mayer-Schonberger points out the importance of forgetting and the risks inherent in the perfect memory we're establishing with digital platforms. *"As we undermine biological forgetting through the use of digital memory we make ourselves vulnerable to indecision or incorrect judgement. This is the curse of digital remembering. It goes much beyond the confines of shifts in information power, and to the heart of our ability as humans to act in time."*[6]

To illustrate his point of view he quotes a wonderful passage from a book by Borges, where the protagonist, Funes, can't forget. He remembers every detail of every moment in his life. *"Forever caught in his memories, unable to think"*, Borges claims, *"to think, is to ignore (or forget) differences, to generalise, to abstract. Since his accident, Funes is condemned to see only the trees, and never the forest. In the world of Funes there is nothing but particulars."*[7]

I guess we all feel a little like Funes, forever sourcing the *'facts'* that are priority-listed in the search engine results pages, unable to get a wider vision or empathise with another person's viewpoint and unable to move on from our past experiences, captured in high resolution and stored forever.

In the recent past the sheer cost of storage would have prohibited this wholesale capturing of every element of our lives. The photos we would have taken, stored in a family album, placed on a shelf and viewed occasionally would have been narrated by a family elder, tasked with explaining the context and adding extra meaning, sometimes to the embarrassment of a son or daughter as their seven year old self is revealed to a new girl or boyfriend. These exchanges were meant lovingly, and fade quickly when turning to the next page of the family photo album. Now they last and last. This traditional, more limited access helped to protect against unwanted distribution and safeguarded the content from being misunderstood or misused. Our new digital world has broken these boundaries in exchange for the effortless sharing and infinite gigabytes of storage available for us to fill up. Data storage costs so little that we rarely bother these days editing the photos we take. We simply snap and share.

Mayer-Schönberger, puts it like this: *"With such an abundance of cheap storage it is simply no longer economic to even decide whether to remember or forget. The three seconds it takes to choose has become too expensive for people to use."* At the current rate of $0.043 per giga byte, according to www.statisticbrain.com,[8] it's hardly worth bothering to be discerning. The risk with such an enormous amount of accessible images, permanently available to anyone with a search engine, is that everyone gets to see aspects of our lives that were previously private or too remote to share. The way we manage these social changes will map our future personally, socially and globally.

As we grapple with the consequences of unlimited access to images and content, we will need to redefine our behaviours,

Picture of patient's living room, published with his permission, via http://casereports.bmj.com/

their impact and what we actually consider to be 'normal'. This was highlighted by the fairly recent diagnosis of an established condition in mental health in this new digital context, where the world's first acknowledged digital hoarder was described in the *British Medical Journal*. The report, published in September 2015, discussed a forty seven year old man who was referred to an outpatient clinic in Holland. The man took as many as a thousand photos a day and spent hours organising them. This interfered severely with his ability to lead a normal life.[9]

Because this was the first acknowledged case of digital hoarding, the clinicians involved needed to test the case against the types of behaviour that were normally associated with physical hoarding. The patient had been previously diagnosed with other associated disorders, for example autism spectrum disorder (ASD) and attention deficit disorder (ADD), and had a history of depression - all of which are common in obsessive behaviour. He had hoarded physical items such as paperwork and bike parts in the past in the belief that they would eventually come in useful. But they soon started to fill his house to the point where he couldn't accept visitors because of the sheer volume of clutter.

The patient's problems with digital started with the purchase of a digital camera five years previously. In an effort to store all the images he was compelled to take, he purchased multiple hard drives and backups. He had real difficulty deleting images, despite regularly reviewing them and knowing that many of them were actually identical, believing they would all be useful some day.

At the risk of sounding crass, there might well be plenty of us who can relate to this story, not all of which is about abnormal behaviour. Many of us take far more photos than we actually look at. Few of us edit them down as much as we should. Lots of us have back-ups, and even back-ups of hard drives. Tools like Dropbox are just one way to stash enormous amounts of data. But this does beg the question: are governing bodies, law, science, and health experts - to name but a few - keeping up with the changing pace of digital and the new definition of acceptable behaviour? What we consider to be rude, invasive, aggressive, peculiar or generally unacceptable in the real world isn't restricted in the same way online, whether it's a behaviour exhibited by another person or a company.

There are 'friends' and there are friends

Old fashioned, traditional behaviour, the type we were happy to share with personal friends and families rather than the Facebook sort,

Lindsey Stone at Arlington Cemetery 2012. Photograph: © Jamie Schuh

previously had to be controlled. Knowing where your images will end up, and what reception they might provoke, is hard when control is no longer explicit now that our audiences are impossibly wide.

Lindsey Stone a charity worker from the US, experienced this after a day out at Arlington Cemetery in 2012.[10] While walking around the Tomb of the Unknown Soldier she played a visual joke, and her co-worker photographed it. This simple act would have previously been of little consequence, but once she shared the image online its context left her control forever. And it proved life-changing. Within just a few hours an audience which disapproved of the image organised themselves into a campaign to get her fired from her job. The company she worked for, Life, (Living Independently Forever), was targeted, and the Facebook page 'Fire Lindsey Stone' quickly gathered 19,000 likes. It took years for her to get over the experience. She didn't leave her house for more than a year and remains terrified that she will lose her current care job with autistic children if the campaign starts again.

The social networks we have created online are often misrepresented as metaphors for the real world, presenting quick, simplistic functions such as 'Like' and 'Friends' as equal to real world rich, meaningful feelings of liking and loving. Our well-founded offline behaviours have not been translated easily to coherent online ones. Sherry Turkle, a cultural analyst and enquirer into the way we live online was once a big champion of the internet. She is now questioning the supposedly endless benefits of digital communication, focusing her studies on the changes to human identity brought about by the internet. *"The social media we encounter on a daily basis are confronting us with a moment of temptation. Drawn by the illusion of companionship without the demands of intimacy, we confuse postings and online sharing with authentic communication. We are drawn to sacrifice conversation for mere connection."*[11]

Notably, a study by Miller McPherson about social isolation between 1984 and 2005 found that the mean number of confidantes of an average person has decreased from 2.94 to 2.08.[12] This is a considerable drop in the numbers of those we often define as our closest friends, which makes this quote by sociologist Eric Klinenberg a whole lot more pertinent: *"it's the quality of your social interactions, not the quantity, that define loneliness"*.[13]

We are all guilty of trying to 'game' friendship, endorsements and fame through social media. Many people in the younger generation have little to compare today's world with, having grown up in a digitally connected landscape where friendships are mimicked between digital and real. As people seek recognition while attempting to navigate the lack of endings, infinite memory and lack of control, we're starting to see problems.

Take NHS Digital, the research arm of the UK's National Health Service, which has published one of many reports showing an increase in mental health issues. One of the most worrying reported a steady increase in women experiencing common mental disorders from 20.4% in 2000, 21.5% in 2007 to 23.1% in 2014. Many commentators associated the phenomenon with increased scrutinising of 'beauty' through shared images, particularly 'selfies'. At the same time the Scottish Health Survey (Knudsen 2016) and other research teams including Hawton and Harriss 2008 and NHS Digital, revealed a growing gap between young women and men who self-harm. As the NHS Digital team pointed out, *"If there is an upward trend in self-harming, with a particularly high rate in young women, there needs to be greater understanding of what is driving this. Some cite bullying on social media as one influence (Daine et al 2013), other sources highlight low self-esteem and anxiety (The Children's Society 2016)."*[14]

Senior research officer Lauren Chakkalackal from the Mental Health Foundation points out that, *"This is the first cohort to come of age in social media ubiquity. This is the context they are coming into and it warrants further investigation. When we are looking at social media and the 'selfie culture' the problem starts far earlier, and I think we are going to see these trends continuing."*

While we can't attribute these changes in mental health conclusively to the use of social media, there are some clear - if not blatant - areas of improvement required around individual identity online, and around protecting your reputation and privacy long term. Balancing the overwhelming social pressure, the frequency of 'share' buttons with opportunities to control, delete and end digital experiences would be a good start.

Who knows who I am?

For businesses in the online world there are significant challenges inherent in the way people share their identity. Consumers in transactions are usually required by businesses to supply a lot of potentially sensitive personal information, which makes many of us suspicious about what is happening to it. This might result in some consumers restricting access to their personal information, withholding, or even lying about it. There is no doubt that there is a general increase in suspicion and even paranoia about what and whom consumers trust.

Pushed to paranoia

Research conducted by Ipsos on behalf of TRUSTe / NCSA, found that 92% of US internet users worry about their privacy online. And 45% are more worried about their online privacy than a year ago. It seems this is limiting business too, with 74% of respondents saying they have restricted their online activity during the last year due to privacy concerns. Forty four percent have withheld personal information, 36% have stopped using a website, and a further 29% have stopped using an app because of privacy worries, of whom 47% claimed this was because they were asked to provide too much information.[15]

A recent anecdote from someone at a major online company revealed that their data appears to be heavily skewed towards older age groups. Apparently an enormous number of their customers are aged a hundred or more. This seems pretty weird for a hip online service? The company believe it's because a growing number of people don't want to give away their true date of birth when signing-up, so they put in any old date. The quickest dates to scroll to at the top of the list happen to be the oldest dates, 1910, 1911 and so on. For many of our consumer interactions online, we'd be forgiven for feeling we're being watched, never really knowing which companies we signed up to in the past or what cookies we agreed to.

In response to this we see more people taking privacy into their own hands. Frustrated by the lack of transparency and protection they receive from service providers, knowing how likely it is to have their browser activity tracked, people have started to use personal Virtual Private Networks – VPNs – software that hides the identity of your computer, making it look like it's in another country. This places control back firmly in the user's hands - although it has a slightly dubious legal footing sometimes. A recent article by *Wired* outlined the world's increasing use of VPNs: *"two-fifths of Indonesians are invisible. More than a third of Vietnamese and Saudi Arabians don't show up either. Take a look at web analytics for Sweden, and you're led to conclude its population is bolstered by large numbers of Netflix-loving ghosts."* Jason Mander, Head of Trends at GlobalWebIndex, said that *"People consider VPNs to be a niche tool, but they are surprisingly widespread. As many as one in four people use them."*[16]

Consensual sharing and shaming

As the media developed and progressed, so did the depiction of porn. We saw pictures of lovers on ancient pottery, explicit engravings in the 16th century, and as soon as photography was invented it too was used for pornography. With the distribution of mass market cameras, pornographic images became more personal and homemade. American GIs going to war would make pinup pictures of their wives to take with them, a highly personal memento to make them feel at home while overseas. Later, pornography magazines searching for new types of content experimented with home-made sections like the infamous 'Readers' Wives'. In the 1980s *Hustler* was one of the first to introduce this, with a section called 'Beaver Hunt'.[17] The feature showcased home made images of consenting couples – or at least that was the intention. These amateur images tapped into the excitement of the moment, and the promises about the imagery 'never being shared' was sometimes short lived as well as unrealistic. Several women sued *Hustler* for publishing their personal photos, claiming the consent forms had been forged and the images submitted without their approval.

Unsurprisingly the internet quickly adapted to pornography, and all the content from the paper-based porn industry came along with it, including home-made images. Sadly the privacy problems inherent in the format also followed, and a stream of non-consented submissions emerged. These tended to be from jilted ex-boy friends. As a result non-consensual sharing is now referred to as 'Revenge Porn', and a number of dedicated websites have been set up to share these images.

A survey and accompanying report, jointly published by the Data Society and the Centre for Innovative Public Health Research,[18] outlines the extent of this worrying trend, citing *"One in ten young women have been threatened with the possibility of public posting of explicit images among all online Americans, 3% have had someone threaten to post nearly nude or nude photos or videos of them online to hurt or embarrass them."*[19]

The legal sector and the industry itself have been slow to react. Many victims have had to embark on often costly legislation to remove unauthorised images themselves, harnessing statute law like the 1986 Computer Fraud and Abuse Act (CFAA), usually used for hacking, and the 1998 Digital Millennium Copyright Act (DMCA), normally used by commercial photographers to get images removed. Despite that, Amanda Levendowski, a Law Graduate from New York who specialises in copyright law, pointed out in a recent interview that, *"80% of all naked pictures found on revenge porn sites are "selfies," which means that copyright belongs to the person pictured."*[20]

Thankfully, given the fast-increasing public awareness of the issue, things are changing. During the last three years more than thirty US states have passed legislation to directly target the Revenge-Porn phenomenon. A national law for the whole of the US has yet to be defined, but a 2016 bill introduced by Representative Jackie Speier aims to criminalise it nationwide.[21]

As consumers of digital services, be they social media like Facebook and Twitter or niche services like porn sites and even providers of storage like Apple's iCloud service – often cited as the place hacked images are stored - we are expected to provide our identities and content, and share it freely on these platforms. The endless features, buttons and prompts found in these interfaces are pushing us to share, and pushing us hard. In stark contrast these groups and organisations often wash their hands of un-sharing, something that commonly falls to the victims, who have to seek their own legal support to get justice.

All this hopelessly fails to maintain a fair balance between on-boarding and off-boarding. We deserve a good balance between the benefits of using a service in the first place and proper support when it goes wrong. Surely we are within our rights to expect more from providers when we want to stop the sharing?

You're done when we say so

An important aspect of the discussion around data is the agreement embedded in the Terms and Conditions, the T&Cs. Looking closer at the tone and intention of T&Cs, it's clear that most fail to balance the on-boarding and off-boarding stages. A company can easily distance itself from any responsibility if the relationship goes wrong, simply by the nature of the legal agreement made in the T&Cs, but the consumer is given little opportunity or support to distance themselves from the company. It's very one-sided. And it doesn't help that almost every set of T&Cs is written in legalese, peppered with impenetrable jargon despite the fact that simple, easy to understand contracts are required by law in many countries. A Symantec survey found that, on average, just 25% of people read T&Cs. The highest proportion is found in Italy, where a surprising 53% of people read them in full.[22] The lowest was in Denmark. Despite many of us clicking 'Agree' when confronted with them, T&Cs are not serving their purpose – to be read and understood. If it related to an aspect of the interface, or a new feature, this horribly poor quality of communication would be considered a total failure.

Getting our fine tooth comb out, let's look at how some companies end the customer relationship through the use of T&Cs, and examine whether it's an amicable or biased process.

Apple's T&Cs miss out on an important opportunity to delight consumers

Apple's iTunes T&Cs, published on the 30th of June 2015[23], contain 20,000 words[24] that the consumer must agree to before using the service. That's pretty daunting in terms of barrier-to-entry, a metaphorical Great Wall of China. Looking through this vast tomb of text, we can see many scenarios where the Apple corporation is entitled to bring their T&C agreement to an end. Just one of these is the use of the word 'Termination' and its context. It would be better to see an example of both parties being empowered to do this, but sadly we don't. 'Termination' provides twelve potential results in various sections. Here are just a few of them:

To terminate the family, in relation to family sharing (I love how dark that phrase is): *"Apple reserves the right to disband a Family in accordance with the "Termination" section of this Agreement."*

In relation to using non-Apple branded products: *"As a condition to accessing your Account or a Service on a non-Apple-branded device or computer, you agree to all relevant terms and conditions found in this Agreement, including, without limitation, all requirements for use of an Account or Service, limitations on use, availability, disclaimers of warranties, rules regarding your content and conduct, and termination."*

In relation to submissions to the iTunes Service, repeated in the App and Book services, it's clear that Apple has the right... but not you!

"Apple has the right, but not the obligation, to monitor any materials submitted by you or otherwise available on the Apple Music Service, to investigate any reported or apparent violation of this Agreement, and to take any action that Apple in its sole discretion deems appropriate, including, without limitation, termination hereunder or under Apple's Copyright Policy."

If you fail, or Apple suspects that you have failed:

"TERMINATION- If you fail, or Apple suspects that you have failed, to comply with any of the provisions of this Agreement, Apple, at its sole discretion, without notice to you, and without waiving your liability for all amounts due under your Account, may: (i) terminate this Agreement and/ or your Account; and/or (ii) terminate the license to the software; and/or (iii) preclude access to the Apple Music Service (or any part thereof)."

And 'termination' as repeated in the App and Book services:

"1. c. Termination. The license is effective until terminated by you or Licensor. Your rights under this license will terminate automatically without notice from the Licensor if you fail to comply with any term(s) of this license. Upon termination of the license, you shall cease all use of the Licensed Application and destroy all copies, full or partial, of the Licensed Application."

Although it says *'terminated by you or the licensor'*, the text only describes how the licensor can end it, not you the consumer. It also mentions that you will *"…destroy all copies, full or partial, of the Licensed Application"*, yet Apple fails to acknowledge how this might be done by you in an amicable manner (or not).

Other variants of the weasel-words around potential endings between the provider and the consumer also exclude an amicable ending. *'Opt out'* gets two mentions, one for the Genius feature in the app and another for the Near Me feature, but neither provides a balanced closure experience.

The acknowledgement of an end date within the T&Cs contains six usages of the word *'Until'*. Most of them refer to you, the consumer, breaking automatic renewal. Other modes of closure include a time-out option of 14 days' refund on some media items you might purchase, but this only applies to iTunes Match, not to the wider service you've signed up for.

Considering all this, it's both revealing and saddening that there are so few options for the consumer to exit from the relationship in an amicable let alone enjoyable way. Apple have hundreds of ways that they can end the relationship, established in their T&Cs, but you and I don't seem to have any at all. In a world where the consumer is supposed to be king, Apple have lost their way. And they're missing a trick too—imagine how much better consumers might feel about Apple is they were given the same basic closure rights as Apple itself.

Facebook T&Cs seem to be improving slowly

Some digital service providers acknowledge the importance of the end of a user experience for their customers, something that creates a decent, fair balance between provider and consumer. Facebook's terms and conditions as of January 30th, 2015 show an improved approach to the legalities. The language is less austere and it talks to the user as an equal, not a threat. It even comes across as nurturing and educational at some points.

Quite early on, Facebook acknowledges that you might want to end your relationship with them. The result is an empowering statement that other providers might learn from. In this reference to IP content, they make an important point about the deleting of content, and how limited they are in controlling it outside Facebook after you've shared it.

"This IP License ends when you delete your IP content or your account unless your content has been shared with others, and they have not deleted it."

In the following passage, Facebook acknowledges your capability to end the relationship by deleting your account or disabling the app. They also acknowledge lingering content in the back-ups that they keep.

"2.2. When you delete IP content, it is deleted in a manner similar to emptying the recycle bin on a computer. However, you understand that removed content may persist in backup copies for a reasonable period of time (but will not be available to others)."

In their Data Policy Facebook shows themselves to be very friendly, clearly acknowledging your capability to delete your account and control what happens after that. This is a very welcome approach, embracing the two-way nature of any worthwhile agreement and valuing your role in it.

Choosing to be open-handed

The Guardian newspaper has taken a different approach with their T&Cs. Published in 2010, they at least feel more even-handed in favour of the user, with a nicely balanced reference to Termination which seems to present the nature of the relationship as two sided.

> "2. Termination of registration
>
> If you no longer wish to have a registered account, you may terminate your account by sending an email to userhelp@theguardian.com. If you no longer accept these terms and conditions, or any future modification to these terms and conditions, you must cease using the Guardian Site. Continued use of the Guardian Site indicates your continued acceptance of these terms and conditions.[25]
>
> If, for any reason, we believe that you have not complied with these terms and conditions, we may, at our sole discretion, cancel your access to the registration areas of Guardian Site immediately and without prior notice.
>
> We may terminate your registered account, at our sole discretion, by emailing you at the address you have registered stating that the agreement has terminated."

The Guardian acknowledges that the user might want to end the relationship, either through cancelling their account or not continuing to agree with the terms and conditions. It carries on to say how this might be done via an email to the help desk.

If more businesses used accessible, less jargon-riddled, plain language for their T&Cs, we might actually bother to read them. We might also feel valued and nurtured through them, since they'd instil a sense of partnership that in turn nurtures a responsibility in the consumer about their own data. In turn this might encourage the consumer to learn more about it and control it appropriately. Facebook alludes to aspects of this, but it's far from perfect.

If we continue with corporate-biased T&Cs which only allow a consumer to leave when they actually break the agreement, we might just usher in a generation of unwittingly criminalised consumers, who have been judged on the way they broke a service agreement. This could bring into effect a black market of service agreements similar to the financial market's bad credit ratings. This would be an unwelcome consequence of our not dealing with T&Cs, failing to make sure that they're fairly balanced between on-boarding and off-boarding.

The Right to be Forgotten

Denying consumers an appropriate ending - a satisfactory closure experience - also fuels debate about the European law of 'Right to be Forgotten' and highlights the lingering nature of content online. Many of us labour under the assumption that the big digital service companies - Google, Facebook and Altaba (previously Yahoo) for example - control the internet and its content entirely. But dig deeper and things are a little more complicated than that.

The issue of content control, particularly in the European Union, has been discussed frequently in recent years. A piece of legislation put forward by the European Union requests the 'Right to be Forgotten' online.[26] The principle has actually been around for a while in varying forms. Britain established the 'Rehabilitation of Offenders Act' in 1974,[27] a piece of statute legislation that requires some criminal convictions to be ignored after the offender has experienced a period of rehabilitation. This aims to remove the criminal record for minor offences that could otherwise stop the individual pursuing a normal life, buying insurance or getting a job. A similar judgement was made in a German court over an ex-convict's wishes to stay anonymous after serving his prison sentence. The lawyers involved cited a 1973 German Federal Constitutional Court decision that permitted the suppression of an individual's name after their sentence was over. More recently France introduced the Right to Oblivion, which sought a legal framework to deliver some control over the way ex-criminals are perceived online.

Further still, the European Data Protection Directive of 1995 aims to regulate the way personal data is processed.

All of the above have provided some legal foundations for the way we find information online and the emerging culture of sharing infinitely between ourselves. It has changed the dynamic of the issue from a niche case for ex-cons to a universal norm that affects us all. We don't, however, enjoy universal agreement about the principles. There's a wide cultural gap between the European Union and the United States, for example. Where many European countries consider the rights of individuals as more important, the United States tends to favour the 'Right of Free Speech' over protecting an individual's privacy. The cutting edge of the debate seems to be legislation established in the 'Right to be Forgotten'. This debate reflects a changing landscape for us all and how we represent ourselves in public, one which highlights the struggle we undergo moving from traditional to digital media.

The recent history of Right to be Forgotten is linked with a Spanish case in 2010, where an individual challenged information indexed on Google referring to his repossessed home. The information was out of date, irrelevant since the issues had all been resolved. The individual believed that the information held on Google infringed his privacy and impacted his reputation negatively. He wanted the references to be removed from the search results on Google Spain and Google Inc. The case was referred to the Court of Justice of the European Union and required judgement on three key issues.[28]

Whether the EU's 1995 Data Protection Directive applied to search engines such as Google;

Whether EU law (the Directive) applied to Google Spain, given that the company's data processing server was in the United States;

Whether an individual has the right to request that his or her personal data be removed from accessibility via a search engine (the 'right to be forgotten').

The Court of Justice ruled on the 13th of May 2014 that:

On the territoriality of EU rules : Even if the physical server of a company processing data is located outside Europe, EU rules apply to search engine operators if they have a branch or a subsidiary in a Member State which promotes the selling of advertising space offered by the search engine;

On the applicability of EU data protection rules to a search engine : Search engines are controllers of personal data. Google can therefore not escape its responsibilities before European law when handling personal data by saying it is a search engine. EU data protection law applies and so does the right to be forgotten.

On the *"Right to be Forgotten"* : Individuals have the right - under certain conditions - to ask search engines to remove links with personal information about them. This applies where the information is inaccurate, inadequate, irrelevant or excessive for the purposes of the data processing. The court found that in this particular case the interference with a person's right to data protection could not be justified merely by the economic interest of the search engine. At the same time, the Court explicitly clarified that the right to be forgotten is not absolute but will always need to be balanced against other fundamental rights, such as the freedom of expression and of the media. A case-by-case assessment is needed considering the type of information in question, its sensitivity for the individual's private life and the interest of the public in having access to that information. The role the person requesting the deletion plays in public life might also be relevant.

The European Union was robust and clear in its directive. Some argue that this is fundamentally a new rule, but the EU believe it is an enhancement of what was already there. It is fair to say that the decision is at least a clarification of an existing law in relation to an ever changing digital landscape, or as the EU put it 'updated and clarified for the digital age'.[29]

The principle behind the ruling supports what we're looking

for with closure experiences in digital services: the ability to remove something that has been put online, to bring something to an end. The Regulation went further, reinforcing the right of the individual and flipping the burden of proof, placing expectations on the company and not the individual. This obviously isn't the case for 'Revenge Porn' or for providers of services that have been hacked.

A more cautious sharer

Despite Facebook continuing to grow at a steady pace, hitting 1.04 billion daily active users in the last quarter of 2015, up 2.97% from the previous year[30], user behaviour is starting to change. According to 'The Information', a digital news site, people are posting fewer personal items on Facebook than ever, a decline that started between 2014-2015, with overall sharing dropping 5.5 percent. The more revealing problem, however, is that people are sharing less about their personal life, something that fell 21 percent over the same period. Apparently Facebook has even assembled a team to focus on the issue.

Pundits are citing many issues that might have created the problem. Maybe the flip-flopping over privacy for the last few years has made people nervous - is it public? Who can see it? What defines 'public'? It could be down to the increase in advertising, or the more savvy news organisations integrating with Facebook to make sharing content even easier. However, one theory seems to be more plausible, and it is directly related to the inevitable consequences of endless sharing and infinite data storage: people are desperately seeking conclusive control and a real ending.

Mary Meeker, in her 2016 *Internet Trends Report*, highlights a shift from the broad social tools like Facebook and Twitter to highly-focused messaging apps like WhatsApp, Facebook Messenger, and Snapchat. Growth across this sector is massive. From 2011 to 2015 it grew significantly faster than regular social media, with WhatsApp alone achieving 1 billion active users per month.[31] Snapchat has attracted 25 percent more users than Instagram in the year to 2016, right across the world in US, UK, Brazil, Germany and France, where it is growing between 63 to 76 percent.[32]

Are we seeing people shunning the mass communication model when interacting with their real friends? We know people adopt different personas with different friend groups - the people you grew up with may know a different you from the people at your work, or the football club you go to. Are people starting to realise the emotional restrictions that come with the mass communication model of social platforms? In a recent interview for *Bitly*, the social marketeer Jordon Scheltgen, from Cave Social, pointed out the importance of authenticity in social platforms: *"The thing that has ramped up Snapchat, in terms of consumption is the lack of automation available on those platforms. You can't automate authenticity, and authenticity is what wins in social."*[33]

Echoing this, Cathy Boyel, *eMarketer's* Principal Analyst says, *"Snapchat has tapped into a key change in consumer behaviour: The desire for intimate one-to-one or one-to-few communication as opposed to broadcast-style sharing across an entire network. This desire is particularly strong among millennials and younger consumers who don't have strong ties to the traditional social networks."*[34]

She goes on to outline the inherent control the user has, and the attraction to its short life span on Snapchat. *"What makes Snapchat different from other mobile messaging apps—and more-established social networks—is the short-lived nature of the messages, the highly visual interface and the features that enable users to get creative with the images they share, and tailor them to specific locations or events."*[35]

The short lived nature of Snapchat seems to have been lacking on other platforms. It allows people to be more emotive, experimental and authentic with just a few people or one person. This is comforting to Snapchat users. Closure is only moments away. The image, conversation or point of view won't last forever, with the whole globe watching. It won't result in the horrors Lindsey Stone suffered, or the public humiliation that victims of revenge porn endure. Snapchat is limited in its permanency, and because of that it mimics our 'real' non-digital conversational life. It's forgiving, it's authentic and it's momentary, just like real life person-to-person interactions.

As digital establishes itself long term, we will experience more of the long term issues other industries have experienced in the past, like the product industry grappling with pollution, and waste – in this sense, off-boarding will become an important aspect of the digital experience. It will tidy the messy aftermath we leave behind, avoiding the risky long term consequences that we see emerging with revenge porn and mental health. We will see a better digital industry which will explore these issues seriously instead of hiding behind T&Cs. We will see empowered consumers able to remove their content as easily as it was to upload it in the first place. We will see the long term security of personal reputation become a vital consideration when people are choosing a provider. In the short term we should see an increase in the use of 'hacks' by consumers wanting to protect themselves, things like personal VPNs. We will no doubt legislate more, discuss more and fail more before improving the wider consumer experience, rebalancing the unfair bias between on-boarding and off-boarding. As Sherry Turkle points out *"Digital technology is still in its infancy and there is ample time for us to reshape how we build it and use it."*

Early 20th century shop with a Credit Card machine.
Skansen. Stockholm

Chapter 11

Business oversight

When I speak to business leaders about improving the end stages of their product or service, they often misunderstand my message as one that would encourage consumers to leave.

Within this belief are two levels of denial. First, that people stay with their providers forever. And secondly, that a person leaving their business should experience something unconsidered and substandard to the rest of the product or service experience.

What I endorse in this book, and hopefully those who have read this far will understand this, is that far from being negative for business, creating better endings would be good for business. It would help it to focus efforts on delivering strategy, broadening and improving long term relationships with customers, and gaining insight and knowledge about their offering.

A big industry of on-boarding.
Can there be an industry of off-boarding?

Once upon a time, products and purchases were very different. People bought only those things which they actually needed, rather than goods which were advertised, or marketed. Customers had very little choice about what they bought. The consumer champion Eirlys Roberts defined this emotionless landscape in her book **Consumers**. *"In the beginning, consumers were not much bothered about the question of quality. They could judge it for themselves. They could tell whether a horse was a horse and not a mule, and even judge its age and speed and staying power, its temper and pride. They could tell silk from wool and good wool from shoddy, a well-baked cake from a sad one, ripe fruit from green, a full-bodied wine from vinegar, and judge whether or not a chair would sustain their grandfather's weight."*[1]

Since then our consumer life has changed a great deal. Industrialisation has generated so many products that it has been necessary to create desire for them, so as to persuade people to buy. Advertisers have built stories around their products and created emotions which impel consumers to make purchases. Marketing has developed techniques around packaging, advertising and promotion to create meaning. Marketing is now a mature, diverse and sophisticated industry.

In response, consumers have been forced to become educated buyers. They have been obliged to acquire, retain and practise amazing skills to enable them to navigate round this complex landscape. If you intend to make a rational purchase, you need to assess and make a judgement about the features of a product and its suitability for your needs, steer your way through the advertising jungle and locate the right purchase for you. Once you've done all this, you may then make an emotional commitment to what you've chosen. And if you don't have the time, energy, expertise or money to buy exactly what you've worked out that you want, then you could feel very disappointed.

We are no longer starved of choice. The impact of abundance has been to confuse us, to offer so many different options that it has become very difficult to off-board consumption, leading perhaps to problems like the obesity crisis, plastic pollution and even global warming.

Society simply doesn't seem able to deal with the consequences of consumption. The business world needs help to off-board our consumption appropriately and safely for the good of society in general.

Might it be possible to develop an industry that adds emotion and meaning to the off-boarding of products and services? Such an industry would help us to enrich and empower the discussion around this issue. Once upon a time there was little on-boarding for products and services. Would it be such a stretch of the imagination to conceive of a world where off-boarding was similarly important and meaningful?

If you're not gonna end it someone else will

The last decade was littered with poorly closed service relationships. Examples from the UK include the mis-selling of insurance in the financial services industry and the failure of the energy companies to inform consumers of the best prices. Such examples demonstrate that customer relationships may not be ended in an amicable manner. The problem is that for many decades it has been accepted that business operates one-way only. So it comes as little surprise that big business is paralysed when it needs to think about good quality consumer experiences at off-boarding and endings.

Two UK regulatory watchdogs have carried out investigations which revealed that energy suppliers and the banking industry have been letting their consumers down, and have proposed very similar solutions. The energy industry has 28 million domestic electricity customers in the UK, and a further 23 million gas customers. The industry have had record number of complaints in the past, seeing a "sixfold increase between 2008 and 2014. Problems related to billing, customer services and payments accounted for the majority of complaints."[2]

The UK energy watchdog Ofgem felt they needed to commission a survey about domestic retail energy, and asked 7000 customers about their service. The findings showed an impressive ignorance of the retail energy market by customers.

Ends.

(a) 36% of respondents either did not think it was possible or did not know if it was possible to change one or more of the following: tariff; payment method; and supplier;
(b) 34% of respondents said they had never considered switching supplier;
(c) 56% of respondents said they had never switched supplier, did not know it was possible or did not know if they had done so; and
(d) 72% said they had never switched tariff with an existing supplier, did not know it was possible, or did not know if they had done so.[3]

Banking has seen similar issues. After the credit crunch of 2008 had rattled trust in the sector, the UK banking regulator - The Independent Commission on Banking - commissioned a report which found very low levels of switching between providers with *"customers switching their current accounts only once every 26 years on average,"*[4] *"Just one in four 'very dissatisfied' customers, and 40% of 'extremely dissatisfied' customers were likely to switch,"*[5] Amazingly *"75% of personal current account holders have never switched".*[6]

The report found many customers felt powerless to switch, believing it was 'complex and risky' and they saw little incentive to do so as there was such a 'poor range of alternatives at other banks', said the report. It was believed this was a result of the way automatic payments are set up - direct debits, standing orders, for example. At that time this took an unconscionable time to change, usually several weeks.[7]

The ICB believed that boosting the ease of switching would help to increase the numbers of people moving to other banks, which stood at 3.8 percent in 2013, when the report was published. It was hoped that this would remove barriers to switching and increase healthy competition between providers.

The banking industry felt the introduction of a maximum transfer time was the best approach and launched the 7 Day Switch Guarantee, in Sept of 2013. The chancellor of the time, George Osborne, believed this was a core pillar to improve the consumer banking industry. He stated that *"The new 7 Day Switching Service is central to our reforms to build a banking system that works for customers".* The service had four expectations of banks that would enable consumers to switch accounts cleanly and efficiently.

1. Giving customers a guarantee that they will be fully protected against any financial loss in the event that a problem occurs during the switch
2. Fully switching over your old account to the new one within 7 working days
3. Allowing you to choose the exact day your account switches
4. Providing a 13 month redirection service, so any debits or credits mistakenly [8]made on the old account are automatically forwarded.

The UK energy industries watchdog made a similar response in 2016. It was inspired by the target of 7 days for the switch that the banking industry had specified, though the switch period was longer at 21 days. *"The Energy Switch Guarantee is a commitment that promises a speedy and safe switch from one energy provider to another. So you won't need to worry about a thing."*[9]

Now in 2017 news is emerging about this initiative, though it's perhaps a bit early to celebrate. There are, however, signs that it is taking off - customer switching has reached a 6 year high.
The Energy Ombudsman, who is the last resort for complaints about energy supplies, has said that customer complaints about switching dropped 36% last year, suggesting companies are also managing the process better. The chief executive of the trade association Energy UK, said the *"high rate of swapping showed competition was working for more and more households"*[10]

As for the banking sector, the success of the 7 Day Switch Guarantee is becoming clear over the last couple of years. BACS, the Bankers' Automated Clearing Services, which oversees payment schemes, clearing and settlement had observed that there had been 1,010,423 switches during 2016. More than 3.5 million switches have taken place since its introduction. [11]

It's concerning that these big industries are in such denial about endings that a third party has to step in to create closure experiences for the consumer. Evidence from the research conducted by the watchdogs was clear that consumers were desperately in need of better closure experiences. Yet the industries could not see the simple imbalance between the on-boarding and the off-boarding for the consumer. The decades of overlooking this aspect of the customer lifecycle by big industries reflects the embedded thinking around the issue. It might be a while before we see some industries taking a lead in off-boarding.

Price comparison sites.

The good news for consumers is that a whole industry has been created to deal with closure experiences independently just for them. Price Comparison Websites, utilise the doubt in people's mind when they are thinking of leaving a service - a behaviour observed by Helen Rose Fuchs-Ebaugh, in her book Becoming An Ex. She found initial doubts are ignited about personal roles as a result of organisational changes, personal burnout, a change in relationships, or the effect of an event. These doubts are then reflected to peers or friends in the form of cuing behaviour. If these cues are recognised by others, then the intention to leave or change the role is reinforced.

Price comparison websites establish doubt in the consumer's mind in a similar way by comparing competing services. If doubts have emerged for the consumer with their current service, this can be clarified by comparison with other services. Comparing services one with another on an impartial basis enables consumers to feel confident about their ability to make a rational choice. They can decide whether to stay with their current service, or whether it will be worth the effort to change.

The second stage of role-exit that Fuchs-Ebaugh observed was the active seeking of new roles. Again price comparison websites provide information for this in consumers. After initially creating doubt in the consumer's mind through comparing prices, they now guide them to the sign-up flow of a new provider. In addition to getting a healthy commission from their activities, price comparison websites have made gains by urging consumers to end their relationships with various providers of energy, insurance, banking etc etc. The companies involved have disregarded the ending of a consumer's relationship with their product or service for many years. Price comparison websites have taken advantage of this and sought to make a quick profit while the companies directly involved have stood by, baffled and indifferent, and missed a great opportunity.

Zombie Apps

The start-up culture of young tech companies seems universally celebrated. Governments love its innovation and energetic drive. Investors love its fairytale prices of soaring equity for their early investments. The industry has many mantras for success and one that is often touted is 'fail fast' which encourages pursuit of quick learning through quick testing. It became such a buzz word, that in 2009 the FailCon[12] conference was established to celebrate the narrative around a journey of business enquiry and failure. The conference has grown, and is now hosted in dozens of cities across 6 continents.

As you would expect these businesses have a high mortality rate. In fact, it is now the same as with new businesses in other industries. The analytics firm CB Insights published a report of start-up failure post-mortems, so know a thing or two about the stories of ending these businesses. They see common patterns of *"companies typically dying around 20 months after their last financing round and after having raised $1.3m"*[13]

Not all of these services simply die. Some linger on for a long time, propped up by the savings of founders who blindly believe that their start-up company is worthwhile. These types of businesses are sometimes called the walking dead. The venture investor and entrepreneur Bruno Bowden diagnosed the psychology behind it as *"the loss aversion principle — the human tendency to strongly prefer avoiding losses to acquiring gains — tilts many towards the former"*[14]

Most start-ups rely on technology in some way. The majority use an app as an interface for their customers. Up to September 2016, apps on Apple's app store that were no longer in use were simply abandoned - they were defined as Zombie Apps by analytics firm Adjust. A Zombie App is an app that doesn't appear in the top three hundred of any of Apple's twenty three different genres lists (including the eighteen different sub genres in games, as well as the categories of Paid and Free). So while there are quite a few opportunities to make it into the top 300, there is also a staggering amount of competition to prevent you.

At the end of 2016 there were over 2 million apps on the App Store and around 100,000 new and updated ones submitted every week. So you can imagine the quantity of apps that don't get seen in your search results will be enormous.[15] Apple have now moved to clear up the App Store and kick out many of the outdated and untended apps. **Tech Crunch** analysis reveals that as many as half the apps had not been updated since May 2015, and 25.6% had been lingering around untouched since November 2013. As of Sept 7th 2016, Apple started to remove apps that hadn't been looked after.[16]

If you are one of the companies needing to remove the app from the app store, you will find little information other than technical instructions. There is no advice on dealing with any existing users, what sort of emotions users might go through at off-boarding or the responsibility the business should have about data on their customers.

This brief instruction comes in stark relief to the voluminous references telling you how to get your app onto the app store, build your customer base and get noticed. So it's no surprise that there are so many zombie apps. Businesses overlook off-boarding, cluttering the digital landscape, and the long term emotional impact to their customers, who will be lingering in limbo with zombie apps.

Design the end in

Far from ignoring the end of the consumer relationship, Kia cars have planned it into their product offering by way of the warranty. Previously, warranties aimed at reassuring the purchaser of a car that the quality of materials was high and the car would last. This would suggest that the ball bearings were gonna last, or the doors were good. These were simplistic messages to the consumer that this was a pretty good car.

All this changed when Kia introduced their Cee'd car in 2007[17]. It was a new approach to the lifecycle of the car - a 7 year warranty. This was a pioneering initiative that shattered the previous norm of 3 years from competitors. Not only is the 7 year Kia warranty a confident endorsement of the company's belief in the quality of its manufacturing, testing and design.

It also introduces a discussion about the end of the car's life, and thus the closure experience. Because Kia has persuaded the customer to think ahead (by promising to look after the car he or she has bought for 7 years) then they are far more likely to have a sensible discussion about the timing for the purchase of a new car, and whether the old one can be repaired, resold or recycled.

This date between Kia and the customer which is seven years in the future, provides a healthy platform for discussion and feedback when the consumer wants to buy a new car and the old one needs recovering, recycling, or re-selling.

Many manufactures don't encourage this open communication channel beyond the 3 years - the length of the warranty. This limits their capacity to recover a product and possibly dismantle it for re-cycling. It also avoids the fantastic opportunity that Kia has embedded into the life of their products - in essence a funeral that brings the opportunity of re-birth (a potential new car sale)

Now in 2016, as the clock ticks around on the first of those guarantees from 2007, it is interesting to note how many of those original Cee'd cars have actually used those warranties. According to the manufacturer, out of the 6160 registered in 2007, the year they launched the 7 year warranty, only 650 have had no warranty claims, meaning 89 per cent have had repairs done to their car under guarantee. [18]Although these are not unusual statistics for repairs of a car over that period, they are certainly re-assuring for customers who want peace of mind.

The honest contract between the consumer and provider around expected life span has obviously done the Kia car business good. The 7 year warranty is a big attraction to new customers. According to COO of Kia Australia, Damien Meredith *"The major reason people buy our product now is the warranty. Price has slipped to third,"* [19]Since introducing the 7 year warranty, the Kia market share has more than doubled, from 1.2 percent in 2007 to 3.2 percent in 2014.[20] Maybe being realistic about the end of life and considering closure can be great for business?

Systematic abandonment

Businesses sometimes ignore the inevitable end of their outdated offering. Peter Drucker, the famous business writer and acknowledged founder of business consultancy, describes how many businesses fail to see the coming end. He lays out this theory in three parts.

First, there are assumptions about the environment of the organisation: society and its structure, the market, the customer, and technology.

Second, there are assumptions about the specific mission of the organisation.

Third, there are assumptions about the core competencies needed to accomplish the organisation's mission.[21]

A business has to excel in all three of these to be a success. Some businesses have managed to keep their business going with a stable set of theories for a long time. But as Drucker points out *"...being human* artefacts, they don't last forever, and, indeed, today they rarely last for very long at all. Eventually every theory of the business becomes obsolete and then invalid."[22]

He points out that many businesses in this situation first become defensive and put their heads in the sand and pretend nothing is happening. After this they try to patch things up and make do. But Drucker believes this never really works. Instead he believes that businesses need to take preventive measures constantly, to prevent their businesses from breaking down.

One of those approaches he calls abandonment.

"Every three years, an organisation should challenge every product, every service, every policy, every distribution channel with the question; If we were not in it already, would we be going into it now? By questioning accepted policies and routines, the organisation forces itself to think about its theory. It forces itself to test assumptions. It forces itself to ask: why didn't this work, even though it looked so promising when we went into it five years ago? Is it because we made a mistake? Is it because we did the wrong thing? Or is it because the right things didn't work?"[23]

These descriptions by Drucker from his 1994 articles in the **Harvard Business Review**, on systematic abandonment, make me ponder the time I spent at Nokia. I saw both the highs - 2007 - 7.2 billion Euros profit and coming lows of 3 billion loss in 2012.[24]

Nokia was stuck in its own perception of the world and its own business theories. They had just fended off the challenge of the Motorola Razr, had record profits and a 50.4 percent market share. Apple had only just released the first generation iPhone in June 2007. Nokia felt comfortable in its market dominance and believed that nothing could challenge their crushing supremacy. They considered Apple to be niche. They considered Blackberry to be a business phone. Their confidence in their direction wasn't backed up with the type of questioning that Drucker champions "If we were not in it already, would we be going into it now?"[25] No-one was heading in to fold-phones in 2012, but Nokia didn't question that logic.

Frank Nuovo was head of design at Nokia until 2007 and largely responsible for the early design direction. In an interview with *The Australian Financial Review*, Nuovo discussed the Nokia mindset, and the Drucker type reasons for its downfall. *"I look back and I think Nokia was just a very big company that started to maintain its position more than innovate for new opportunities. All of these were in front of them and Nokia was working on them, but the key word is a sense of urgency. While things were in play there was a real sense of saying 'we will get to that eventually.' In hindsight it got the balance fatally wrong, servicing its existing product at the expense of fresh innovation."*[26]

This aversion to considering endings is deeply engrained in the culture of business, and in the way we all think. The terror management theory, that I talked about earlier in the book, establishes an aversion to considering endings, even if it is healthy to do so for our businesses. Drucker offers a wonderful method of reckoning forcefully with a business and its potential endings periodically. It widens business thinking, and encourages a healthy, honest relationship with endings. Where there is indifference and cynicism, he encourages action and conscious thought. He promotes thinking about endings.

Please, no more choice

In the earliest of our economies the idea of choice was pretty rare, consumers had to buy what was available. Industrialisation provided consistency of quality in products and a wider range of options for the consumer. The championing of choice has remained strong ever since, and created many innovations that deliver it. The supermarket is a prime example.

Early grocery stores were staff-intensive with items being fetched manually by staff members behind counters. The self-serve type of supermarkets we use today empower the consumer to browse and make choices. First introduced by Clarence Saunders with his Piggly

Wiggly stores in 1916,[27] supermarkets have become a common feature across the world.

Roll forward a hundred years and questions are starting to emerge. Have we exhausted the route of increased choice as a driver to growth? How much choice do we need? And how do we remove the leftovers?

The author Barry Schwartz in his book the *Paradox of Choice* argues that *"eliminating consumer choices can greatly reduce anxiety for shoppers"* and that *"choice no longer liberates, but debilitates. It might even be said to tyrannise."*[28]

In the supermarket landscape of the UK, we are starting to see this in action, as some of the bigger supermarket companies, who champion choice, are starting to get challenged by smaller, more focused retailers.

Britain's biggest supermarket, Tesco, stocks 91 different shampoos, 93 varieties of toothpaste and 115 versions of household cleaner.[29] Despite this being a baffling array of choice it previously made good business sense and provided Tesco with a 28% market share in 2016.[30]

A more recent entrant to the UK supermarket sector is Aldi, which arrived in 1990[31], and has become the UK's favourite brand according to the Brand Index[32]. Lidl, which offers a similar range of limited choices, became the nation's number 2 favourite in the same survey.

Ronny Gottschlich, Managing Director of Lidl, believes a great deal of the success of this type of supermarkets is down to the fact that while the range of goods is relatively limited, they are of high quality. He describes it as *"If you [and another customer] don't know each other, would you like to pay for his choice of a different type of water? If you go to another retailer that has got 20 different types of water, someone has got to pay for that space, someone has got to pay for that rent."*

Graham Ruddick, retail correspondent for *The Telegraph Media Group*, goes as far as to say *"the secret to their business model is based on the fact they sell fewer products than a typical Tesco. While Aldi sells 1,500, a Tesco supermarket can sell 40,000."*[33]

Ends.

It is not just the bigger supermarkets that have been overwhelming consumers with an inordinate number of choices. Proctor & Gamble reduced the Head and Shoulders shampoo product line from 20 to 15 - and saw a 10% increase in sales. [34]

It would be flippant to suggest that Tesco reduce its product line from 40,000, to 1500, but there is obviously a change of heart among consumers, who are becoming less keen on having a huge range of goods to choose from. To make such changes would obviously mean a reduction in the number of goods on sale, and even put an end to some product lines. Although this does sometimes happen, it might not be popular with some of the larger producers and retailers, since it could result in the writing off of large investments.

The time and effort needed to generate emotional traction with consumers through branding and advertising is costly for new product lines. In contrast, the removal of these product lines currently produces little emotional reaction. There will, however, always be consumers who are emotionally attached to some products which are not hitting their target sales, or may even be removed from sale. The question is, how might you utilise the energy from that emotional attachment when the product's life comes to an end? Can you re-direct it? Or do the messages stop, and the producer just falls silent about that product?

Re-thinking this situation would not mean bringing back discontinued products in hopeless nostalgia. It would mean doing more with the emotional energy put into a product's launch or lifespan. It should be possible to create an off-boarding that is of benefit to other products or the brand, transforming the emotional energy generated into something positive rather than just letting it crumble away into silence.

Meaningful goodbyes

The celebration of the 'fail-fast' culture in the world of start-up businesses hides a wider problem for the consumers who support them. These consumers provide vital early support for these start-ups.

Ends.

As customers, they contribute their identity plus access to a whole range of personal data - photos, music preference, location, address book, to name a few.

As these business grow and become valid entities they attract the attention of larger competitors who buy them up. The businesses who don't make it this far go bankrupt. Either way, the start-up changes form and has to reconcile this with its original customer base. It needs to alert them to potential changes in the service offered, which often results in the founders believing that they need to draw some sort of a line under the 'journey' of their business. This often compels them to bid farewell to their supporters and customers by means of a pseudo emotional letter.

Phil Gyford has been studying and sharing these letters for half a decade now on his blog 'Our Incredible Journey'[35]. They reveal a strange practice, particularly in regards to closure experiences.

I met Phil on a crisp April day in 2015, at a coffee shop in London to chat more about the blog and the phenomena of these letters. Phil is a tall imposing presence, with a diverse intellect - as comfortable talking about psychology as he is about databases and start-ups. I wanted to know what made him start the blog, what was the point of capturing these ending moments in the start-up world, and why were they all so similar?

He choose his blog's name from the cliche phrase many of the founders use when writing these type of letters. He said *"I felt compelled after seeing a similar pattern of 'We have sold!' followed by 'We are going to delete all your stuff' and a warm 'Thanks for everything!'"*

He revealed that *"These statements of selling out and deleting your stuff, don't always come at the same time. Some might be separated by months even years - possibly down to a slow technical integration between seller and buyer, but it always comes along at some point. In response the users either have to move their content, transfer files to a new platform or lose the lot."*

Phil wanted to show how many companies are doing this and

make the issue more visible to the start-up community. He had been on the wrong side of the issue once before, losing content and data, due to a service shutting down.

A typical example can be seen on the website, with the announcement from Pebble, the watch manufacturer, from December of 2016. Phil edits the posts down so visitors to his site can see the essence of these announcements from founders.

"... Pebble is no longer able to operate as an independent entity. We have made the tough decision to shut down the company and no longer manufacture Pebble devices. Pebble is no longer promoting, manufacturing, or selling any devices. Pebble devices will continue to work as normal. No immediate changes to the Pebble user experience will happen at this time. Pebble functionality or service quality may be reduced in the future.

Making Awesome Happen will live on at Fitbit. ... many members of Team Pebble will be joining the Fitbit family to continue their work on wearable software platforms." [36]

Another example from Readmill in March 2014:

"We're proud of the product we built, but even more so, we're grateful for the community of readers that made it grow. At every turn, your feedback shaped Readmill's development, and your passion signaled a new chapter for reading. Together we wrote in the margins of ebooks and discussed our favorite passages from across the world. Thank you for helping to bring this reality into view. As of today, it is no longer possible to create a new account, and on July 1, 2014, the Readmill app will no longer be available.Our team will be joining Dropbox, where our expertise in reading, collaboration and syncing across devices finds a fitting home."

Shutting down can be disappointing to many customers who have been passionate about a service. Some may have supported it with their own money, as in the case of Pebble which was an early Kick-Starter success story. They may feel like they have an emotional investment in the start-ups they support, so knowing that the service is being shut down comes as a personal disappointment. The frustration is evident from looking at feedback on the App Store from customers after an announcement letter has been published. This example from Readmill, is typical.

"We can't rely on 3rd party services to carry us. Somewhere down the line money becomes a major issue and the whole idea of establishing a brand goes out the door."

The start-up industry is awash with failing companies. In fact it's sometimes considered a badge of honour by investors who might value the experience of people who have been through the problems of a failing business and learnt what not to do. So endings shouldn't come as a surprise. Sadly, the long established culture of overlooking consumer endings in the wider business community is being adopted by the smaller start-up community. A common denial of endings seems to be becoming a norm.

Sharing the blame with the consumer

Over the last couple of centuries, business has been adopting an increasingly considered approach to society and its problems. At the same time, it has - perhaps unwittingly - reduced consumers' responsibility for consumption

Business previously saw the world as a one dimensional opportunity for profits over everything. The economist, Milton Friedman, proudly championed this cultural pursuit when he said *"There is one and only one social responsibility of business — to use its resources and engage in activities designed to increase its profits.".*[37] As a result of social pressure, and government intervention, we have moved beyond that rhetoric and now have a more universally beneficial business approach.

Business leaders align now more with the Nobel Prize winning economist, Joseph Stiglitz, who said *"Whenever there are externalities— where the actions of an individual have impacts on others for which they do not pay, or for which they are not compensated—markets will not work well."*[38]

Corporate social responsibility

Balancing the needs of business and society are common in today's economy. Many businesses are choosing to run Corporate Social Responsibility (CRS) programmes. These cover many different areas of society: support for social or environmental issues is common, safety at work, going beyond labour laws, announcing environmental initiatives or supporting charities.

There is a great deal of evidence to suggest that for a company to engage in a CRS will have a positive impact on the business, not only on the public perception of its brand, but also on the loyalty that it stimulates among consumers. The most successful approach is to align the CRS with the wider business.[39] For example - a large manufacturing company announces support for local pay rises in its China factory, or an oil company funds a wildlife charity, or a sports company supports a youth sports charity.

Consumers also respond better to a CRS programme if they understand the links between the environmental efforts and their own benefit. A recent study by the Tuck Business School found that *"The message should be concrete about the positive effects of, say, reducing waste or electricity use. When retailers focus on environmentally friendly initiatives that lower their costs and lower prices for consumers, that can really work,"*[40]

Business is blamed for consumer actions

The rationale for the setting up of a CSR programme is to demonstrate that business is involved in society's progress. One of the biggest issues within this is the impact consumption has on the environment. Navigating responsibility around consumption is a difficult subject. Many companies now recognise their role in this and through their CSR, encourage change and improvements to environment damaging processes. However, damaging the environment isn't done by business alone. What needs to be acknowledged is that very often business is operating on behalf of the consumer, providing the means for consumers to damage the environment. And, because business seems to be taking the blame for this damage, then the consumer's part in it is downgraded, unacknowledged. As one party increasingly steps up to acknowledge their role, the other party (the consumer) is distanced into the shadows, hidden from blame.

We need to change this and recognise the responsibility of consumers - and, indeed, ourselves, since we are all consumers. This should be done not via guilt-ridden, passive messages, but by the creation of active off-boarding experiences. By this means we can recognise our responsibilities with achievable actions to counter the problems with the environment.

Examples of actionable endings

Air Travel
We can see a good example of this in the way air tickets were bought a few years ago. At the end of the purchase process, the consumer was provided an opportunity to balance their carbon impact of the flight. This created an awareness of cause and effect in the context of on-boarding and off-boarding. The benefits and emotional triggers that motivated the consumer to buy the ticket were balanced by giving back after the trip through the planting of trees, for example.

This helped many consumers to feel as if they had some actionable control over the damage they were inflicting on the environment. It balanced their consumption. Offsetting our carbon on flights has now been tucked away under the wing of many airlines' CSR programmes which have, in turn, shifted the blame away from consumers.

Ends.

Social Networks

Although not attached to CSR programmes or the environment, a similar problem arises in the digital industry over the matter of sharing. The consumer is deprived of actionable control over the negative impacts of their consumption. When we take a photo, video, or share a news story; encouragement to share at the beginning of the customer lifecycle is not balanced with actionable ability to remove, or un-share thereafter. Numerous prompts to share, in highly visible locations in many interfaces, are not balanced with easy ways to remove the same content.

As a result, consumers have to challenge the providers of services when they need to remove content instead of being empowered to do this themselves.

Car scrapping.

A better example of the consumer experiencing actionable endings can be seen when scrapping a car. Previously a person could take their old car to a scrap yard in return for cash and no questions asked. In recent years, however, this has changed. The End of Life Vehicle Directive[41] now requires the 2 million cars scrapped each year in the UK to be subject to appropriate processes. The owner's identity is attached to the vehicle via the Driver & Vehicle Licensing Agency[42] which handles road tax and driving licenses in the UK. Through the Scrap Metal Dealers Act cash is removed from the process and makes the transactions traceable. Since 2005 the breaking up of the vehicle has to be done at an Authorised Treatment Facility and so cars are broken apart without damaging the environment.[43] If the owner is looking to replace the scrapped car with a new car, manufacturers are now required to take customers' old cars back for for free, even offering a trade-in amount if the car is worth anything. Consumers of cars now have access to a variety of off-boarding methods. Each method has clear, actionable steps for the consumer.

Ends.

Looking again at endings

Talking about endings is difficult, repulsive even. As individuals, as consumers, we have a psychological aversion to it - and businesses are positively afraid of it. Business teaches us to ignore it, to avoid it and to blindly follow the growth and acquisition model. This is a model which has been crafted and updated over centuries. But because of this blindness, many solutions have been overlooked, opportunities have been ignored and different pictures have been missed.

Engaging with the end of the customer lifecycle can yield better discussion with customers, gaining important information which has been overlooked in the current approach. It provides opportunities to take the lead in off-boarding, creating innovation and new relationship methods, before an industry watchdog enforces the acknowledgement of endings in your industry.

It provides a platform to share responsibility with the consumer and partner on some of the biggest problems with consumption. And business shouldn't be allowed to be the sole culprit in the poisonous narrative of consumption.

Engaging with the end of the customer lifecycle can help to create a healthy business by transforming the discussion about the end of product lines, services and outdated technology. It provides a supportive vocabulary for talking about how to end the things we cherish and creates conclusive and healthy ways of challenging businesses which are performing badly.

Far from closure experiences being bad for business, they can be the solution to the most complex of consumer ills.

Conclusion.

I hope that reading this book has convinced you of a number of things. The most fundamental is that you now see the importance of endings in the consumer lifecycle. And I hope that, as a consequence, you will appreciate that society has rejected endings thanks to changes in religious practices and the emotional distancing of death over time.

Also, I hope that I have demonstrated how gradually distancing our links with waste has been effected by consumers no longer having a close, actionable relationship with it. The consequence of this is that we no longer feel responsible for the damages of over-consumption.

This regression has been partnered with the increased pace of consumption, but more worryingly a tethering of identity to consumption of virtually everything - products, services, digital. This means that individuals are identified and celebrated through acts of consumption, and the blame for the damage to, say, the environment, has been ignored and removed from the individual.

This book has attempted to illustrate how we think and feel about endings and encourage us to consider some of the wide ranging work in the field of psychology. This has revealed how our mind frames endings and particularly the repulsion we have with death. We saw how deep and hidden this sometimes is by looking at Terror Management Theory, put forward by Ernest Becker. This theory suggests that, subconsciously, we avoid acknowledging our limited lifespan and that we try to avoid anything to do with it. This has, in turn, helped us to ignore the importance of endings in other spheres of our lives.

Human made narratives - films, books and games - provided examples that emotional endings can be meaningful and important. These examples provide an interesting counterpoint to the consumer world of products, services and digital. Narratives show us that examples of structured, organised off-boarding can encourage emotional, and meaningful experiences.

The emotionally cumbersome world of divorce provided a compelling example of the ingrained contrast between the on-boarding to marriage and the hapless approach we have with divorce. When blame needs to be associated and proven or in some extreme cases an individual can become outcast and immediately destitute after the utterance of 3 words.

We saw how bias had evolved in commercial services and the financial services industry. Who, although having some of the worst examples of closure experiences and consumer endings, have potentially the most to gain from improving their off-boarding. Despite this they have developed a culture of short-termism and denial that endings could be positive or provide improvement to their customers' experiences.

We then reflected upon the world of products, which is dominated by the mounting, daunting avalanche of electronic waste. This area provides many examples of the incredible bias in the messages the consumer receives over the life cycle of the product he or she has bought. At off-boarding, in stark contrast, they are offered little information about what to do with a product after its useful life is over. This results in millions of devices being tucked away in drawers and on shelves in our homes. Consumers really don't know what to do about the poisonous metals in their gadgets and how to protect the personal data in them.

We all had such hopes for the new landscape of digital, but alas, it has inherited the bad habits of other industries in relation to endings. The enormously high turnover of digital products has created a wasteland of lingering private data, which has been captured via the photos, messages, comments that we record every day. The interfaces to this world are awash with encouragement to share, sign-up, create. This encouragement hopelessly overlooks the need to balance out all the inputs against easy-to-action ways of un-sharing, controlling and removing. This is a situation that is slowly undermining our reputations and even our self esteem, most notably in the younger generations using these digital tools.

For the last chapter in this book I wanted to highlight the opportunities businesses tend to overlook due to their cultural opposition to endings and off-boarding. Instead of being a threat, which is the most common misperception, it can be a useful and potent tool for industry. In this chapter I also wanted to highlight the injustice of blaming industry for the ills of consumption. Illogically, industry often seems happy to take on that guilt, hedging it against the risk of upsetting customers. But I argue that the customer often isn't exposed to actionable alternatives to a bad ending. Were industry to provide better consumer experiences at off-boarding, then maybe a more just balance in blame would be present around the ills of consumption.

Future of endings

I hope you see the urgency in changing our approach to the ills of consumption. It's not enough to use the simplistic approaches that we currently use. The trouble with those is that they generate conflicting messages - buy more, then feel guilty. We need to create meaning inside the consumer experience that stimulates action in the context of consumption.

Ends.

The biggest improvement will be if you are now aware of the endings in our consumer life cycles. You will notice their absence, or recognise their presence and quality. Maybe some of you have influence in the product development process as designers, developers, owners or managers.

You will no doubt find it difficult to convince many people of the need to balance the consumer lifecycle. It is a difficult issue to bring up. It goes against a deeply ingrained psyche for much of business culture.

None of the approaches to business, the way we structure our projects, the aims and objectives of the briefs we receive or the projects we run have a consideration of endings built into them.

The business case

To support these difficult discussions with your peers, I have found that one of the best examples to give for business leaders is the story about gym membership from earlier in the book. It provides a tangible case study in business terms and challenges the denial that customers will be yours for the long term. Adopting a good closure experience for that example would create clear benefits to the business in a unique and creative way.

Once your company, business or product development team are listening you will need to suggest a structured approach around the product development effort. This may come in the form of everyone involved promising to be a little bit more aware of the ending, but I wouldn't depend too much on those types of promises. Instead I suggest you place it more formally into the product development process.

Planned endings

An approach I often recommend is the 5x5 approach. This would see five percent of your project effort given to investigating your product five years from now - 5 percent at 5 years out. Once you start considering your product after five years you have to start considering its obsolescence and therefore people off-boarding from it. You will then have some theories and even hypotheses around what needs to improve your current product to help consumers off-board.

You will then need to consider what is quality, and what we are aiming for. What is a good closure experience? In the work that I have been doing on this subject over the last 12 plus years, I believe it comes down to this.

A good Closure Experience should be Consciously Connected to the rest of the experience through Emotional Triggers that are Actionable by the user in a Timely manner.

Let's look at these in some more detail.

Consciously connected

Many of the experiences we have at on-boarding don't have an appropriate counterpoint at the off-boarding stage. This leaves the experience unbalanced and means halting the conclusion of a narrative that we were sold when we purchased the product.

One example I like to use to illustrate this issue is purchasing a Canon printer ink cartridge after my printer ran out of ink. I drove to the local office supply store and purchased a new cartridge, returned home, replaced the old one with the new one and turned on the printer in the hope that all was working. The familiar buzz, buzz, sound came out of it and I breathed a sigh of relief. This turned to bafflement when I looked to the old, now redundant printer ink cartridge I had just replaced. What do I do with it now? It is highly poisonous. I can't put it in landfill. I can't put it in recycling. Ah, I thought, there must be something on the packaging to instruct what is best? But this wasn't the case. Instead there were a couple of marketing messages trying to sell me additional services - more starting experiences.

I turned my search to the computer and found that Canon had been offering a very good recycling, and reclaiming service since the

90's. I went to the Canon website and sure enough there was a service that I could use to recycle old Canon cartridges. After reading through 2300 words of Terms and Conditions, I could finally get a small plastic envelop sent to me, so that I was able to return the old poisonous cartridges to Canon.

The Canon printer ink lifecycle is clearly broken. I have spent significant efforts off-boarding a Canon product myself. Canon clearly have not considered the context of the user through the entire consumer lifecycle. They have only focused on the on-boarding and usage phases of the customer experience. This undermines their good work on recycling and reclaiming their ink cartridges. This is a common mistake in industry, where provision of material recycling is invisible to the consumer.

Amongst the many simple solutions to remedy the Canon printer cartridge problem would be placing return packs in the capacious packaging (there is 500ml of redundant space I measured, yes I am geeky) they sell their product in. This would keep the life-cycle coherent between the on-boarding and off-boarding. It would provide the user an actionable way to recycle and inform the user that Canon is serious about the environment, and that they can create closure experiences, as well as starting ones.

Emotional Triggers

Emotional triggers should be balanced in meaning between the on-boarding and off-boarding. This would be similar to the approach we looked at earlier with regard to making films. The off-boarding of a product or service should emotionally match the way we start and end a relationship.

When I was a kid in the 70s and 80s I watched, as many of us did, our parents smoke. It looked pretty fricken cool. The packaging on these products was some of the best around. Printed with foils and special inks, they had well thought through designs and sophisticated branding. At the bottom of these packs, very subtly, was a little nod to the issue of long term damage as a consequence of smoking. This was a tiny little reminder of potential off-boarding from this product experience.

A little later governments began to put more pressure on the producers to place warnings on the packs. These used a cold hard graphic approach of black text on a white background, which offered a pretty emotionless message in stark contrast to the amazing packaging. But still smoking remained popular. Many people overlooked the health risks as they did the warnings on the packs. In 1991, the European Union introduced laws that required a warning on the front and back of the packs. This was increased in 2003 to requirements of 30% of the surface of the pack. But still the message was in black and white, emotionally barren.

Eventually emotion was introduced to the warnings and in 2008 pictures of patients suffering as a result of smoking were included. This was increased in 2012 and saw Australia enveloping the entire pack of cigarettes with warnings and imagery of the risks of cancer, becoming the one of the first countries to enforce Plain Tobacco Packaging. This removes any branding from the producer - colours, fonts, imagery, and replaces it with a basic sans serif font with verbal and image based warnings.

We can interpret this story as the completion of the 45 year journey of dominant starting experiences / on-boarding messages in the 60s and 70s, which had dominant branding message - designed to

Ends.

inspire positively the image of smoking to the consumer. To a dominant closure experience / off-boarding message in 2012, that only saw negative messages about the products long term use on the packaging.

If the balance of on-boarding and off-boarding in emotional messages was more even and meaningful earlier on in the story of 20th century smoking, maybe consumers would have listened to the messages more. They may have saw the communication as even handed, not clearly from another source, and as part of the same experience.

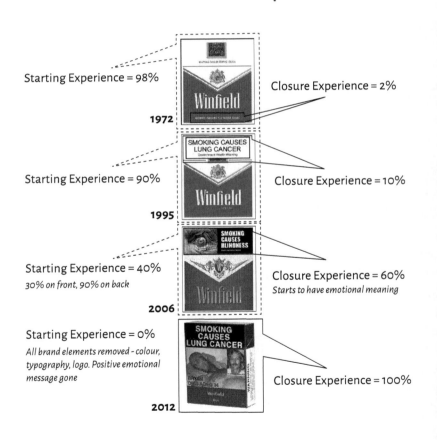

Starting Experience = 98%
1972
Closure Experience = 2%

Starting Experience = 90%
1995
Closure Experience = 10%

Starting Experience = 40%
30% on front, 90% on back
2006
Closure Experience = 60%
Starts to have emotional meaning

Starting Experience = 0%
All brand elements removed - colour, typography, logo. Positive emotional message gone
2012
Closure Experience = 100%

Actionable by the user

The encouragement we receive to sign-up and share with digital products is often overwhelming. In contrast, easy methods to remove items, or end the relationship easily and completely are overlooked. As a consequence consumers are casually creating a lingering legacy from their content and, in the majority of cases, ignoring the risk that content might bring long term. Snapchat and its timer feature is a welcomed counterpoint to this. Initially the timer feature was useful for lovers to send each other private, *"high-risk"* messages that would self destruct, but an increasingly aware consumer base is now seeing the long term benefits of placing an expiry date on their digital content.

Adding an actionable ending, like the timer, is good for consumers. It allows them to take responsibility for their consumption. Too often companies are relied on to be guardians of our actions and our content. Instead companies should empower consumers with actionable endings.

Timely

Many of our consumer relationships can linger on long after their useful life has ended. We see evidence from the clutter in our houses and the growth in the usage of personal storage units. Recognising when is a good time to end a relationship is important. Too long, and we are living in denial that it has ended, to short and the consumer feels cheated. We know that an end is likely to come, and considering what type of preparation is required is a good thing. We need to be realistic about timing.

The last example I want to give is about Do Not Resuscitate (DNR) forms. Many countries have them, or something similar. They are an agreement between a patient and medical staff that, in the event of the patient expiring they will not to be resuscitated. These are usually provided to people who feel their physical quality of life is not to a standard that they feel is worth continuing. Or that they have had so many procedures that they feel exhausted with medical interventions

Ends.

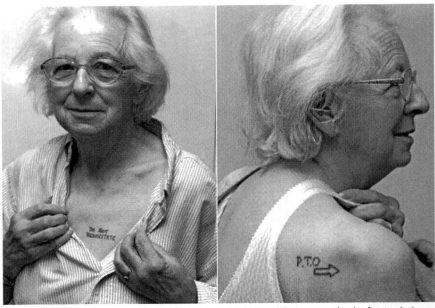

Do Not Resuscitate (DNR) Tattoo. www.deathreferencedesk.org

and instead they want to let nature takes its course.

The forms have often been the cause of controversy. In some cases people have been resuscitated against their wishes, which has had the unexpected result of causing some elderly people to get their first tattoo. Others have argued that DNRs have been used on occasion without consent.

This humble letter size form has certainly caused a storm. Maybe we shouldn't be surprised, compressing the last days on earth into a printed form is the wrong approach. There is obviously a lot to be considered at these times, and much to be discussed between the patient, the doctor and the patient's family. It seems ambitious to place it all into a form, when we rarely discuss death throughout the rest of our lives. Are we equipped to discuss it properly in the fractions of time we might have left? Many aren't prepared. Repelled by the idea of the ending of our lives, we prefer ignorance to acknowledgment.

Sherwin Nuland, the physician and author of the book *How we die*, argues "*It is only by frank discussion of the very details of dying can we best deal with those aspects that frighten us most. It is by knowing the truth and being prepared for it that we rid ourselves of that fear of death that leads to self-deception and delusions.*"[1]

The DNR form examples that delusion. A quick fix at the end of life won't bring the desired dignity in death many would wish for. Death should be discussed in the context of life. Acknowledging it, discussing it, and preparing for it would reveal its nature, and remove some of its horrors.

We should leave this discussion with another physician, an associate of Aristotle, Alkmeon, who said

"*Men die because they cannot join the beginning and the end*".[2]

Ends.

Ends.
notes

Chapter 1.

1. *Mintel. (2017, April, 20)* http://www.mintel.com/global-new-products-database
2. *Read.S. (2013.4.10).* The Independent. A quarter of adults have lost a pension pot, says survey. Retrieved from: http://www.independent.co.uk/money/pensions/a-quarter-of-adults-have-lost-a-pension-pot-says-survey-8567984.html

3. *Read.S. (2013.4.10).* Independent. A quarter of adults have lost a pension pot, says survey. Retrieved from: http://www.independent.co.uk/money/pensions/a-quarter-of-adults-have-lost-a-pension-pot-says-survey-8567984.html
4. *Wikipedia. (2017, April, 13)* Maslow's hierarchy of needs. https://en.wikipedia.org/wiki/Maslow%27s_hierarchy_of_needs

Chapter 2.

1. *McMahon. R. (2011)* On the Origin of Diversity. p. 72.
2. *Kearl. M. C. (1989)* Endings. A sociology of death and dying. p. 35
3. *Sellman W.M. (2014)* Brief History of Death. p. 25
4. *Bode J. and Olyan. S. M. (2012)* Household and Family Religion in Antiquity. p. 66
5. *Sellman W.M. (2014)* Brief History of Death. p. 37
6. *Kearl. M. C. (1989)* Endings. A sociology of death and dying. p. 180
7. *Viralnova (2014.10.12).* What Heaven Looks Like According To Various Religions. Retrieved from URL: http://www.viralnova.com/heaven/
8. *Brahmavamso. A. (1990.6)* What the Buddha Said About Eating Meat. Retrieved from URL: http://www.urbandharma.org/udharma3/meat.html
9. *Religion Facts (2015.3.17).* Jehovah's Witnesses Beliefs. Retrieved from URL: http://www.religionfacts.com/jehovahs-witnesses/beliefs
10. *Mormon.org (2017).* What happens after we die? Retrieved from URL: https://discover.mormon.org/en-us/topics/life-after-death/
11. *Mormon.org (2017).* What happens after we die? Retrieved from URL: https://discover.mormon.org/en-us/topics/life-after-death/

12. *Viralnova (2014.10.12).* What Heaven Looks Like According To Various Religions. Retrieved from URL: http://www.viralnova.com/heaven/
13. *Abdulsalam. M. (2015. Dec.27)* The Pleasures of Paradise. Retrieved from URL: http://www.islamreligion.com/articles/11/pleasures-of-paradise-part-1/
14. *Wellman. J. (2011)* What Does the Bible say Heaven is Like? Retrieved from URL: http://www.whatchristianswanttoknow.com/what-does-the-bible-say-heaven-is-like/
15. *Holy Bible, New International Version (1973)* Revelation 21:4 Retrieved from URL: https://www.biblegateway.com/passage/?search=Revelation+21%3A4
16. *Jayaram. V. (2016).* Death and Afterlife in Hinduism. Retrieved from URL: http://hinduwebsite.com/hinduism/h_death.asp
17. *Wikipedia. (2017.3)* Tithe. Retrieved from URL: https://en.wikipedia.org/wiki/Tithe
18. *Mormon.org. (2017).* What is done with the tithing that Mormons pay? Retrived from URL. https://www.mormon.org.uk/faq/topic/tithing/question/church-tithing

Ends.

Chapter 2 cont...

19. *Pennington. Ken. (2015).* The Black Death and Religious Impact. Retrieved from URL: http://faculty.cua.edu/pennington/churchhistory220/LectureTen/BlackDeath/Religious%20Impact%20page.htm

20. *history.com Staff. (2010).* Black Death. Retrieved from URL: http://www.history.com/topics/black-death

21. *history.com Staff. (2010).* Black Death. Retrieved from URL: http://www.history.com/topics/black-death

22. *Wikipedia. (2017).* Consequences of the Black Death. Retrieved from URL: https://en.wikipedia.org/wiki/Consequences_of_the_Black_Death

23. *Wikipedia. (2017).* Consequences of the Black Death. Retrieved from URL: https://en.wikipedia.org/wiki/Consequences_of_the_Black_Death

24. *Wikipedia. (2017).* Nicolaus Copernicus. Retrieved from URL: https://en.wikipedia.org/wiki/Nicolaus_Copernicus

25. *Wikipedia. (2017).* Reformation Retrieved from URL: https://en.wikipedia.org/wiki/Reformation

26. *Wikipedia. (2017).* Protestant culture.Retrieved from URL:https://en.wikipedia.org/wiki/Protestant_culture

27. *Mitzman. A. (2001.9.28).* Max Weber. Retrieved from URL: http://www.britannica.com/biography/Max-Weber-German-sociologist

28. *Mitzman. A. (2001.9.28).* Max Weber. Retrieved from URL: http://www.britannica.com/biography/Max-Weber-German-sociologist

29. *Beder. S. (2004)* 'Consumerism – an Historical Perspective', Pacific Ecologist 9, pp. 42-48.

30. *Aries. P. (1976).* Western Attitudes towards death from the Middle Ages to the Present. p. 70

31. *Aries. P. (1976).* Western Attitudes towards death from the Middle Ages to the Present. p. 70

Chapter 3

1. *Jefferies. J. (2015)* The UK population: past, present and future. p 2.

2. *Hannan. M.T. and Kranzberg. H. (2017.6).* History of the organisation of work. Retrieved from URL: http://www.britannica.com/topic/history-of-work-organization-648000

3. *Strasser. S. (1998)* Getting and Spending. p 279.

4. *Strasser. S. (1999)* Waste and Want. p 13

5. *Strasser. S. (1999)* Waste and Want. p 6

6. *Dunson, Cheryl L. (1999.12.1)* Waste and Wealth: A 200-Year History of Solid Waste in America, Waste Age.

7. *Wikipedia. (2017).* John Snow. Retrieved from URL: https://en.wikipedia.org/wiki/John_Snow#Cholera

8. *BBC. (2017).* History. John Snow (1813 - 1858) Retrieved from URL:http://www.bbc.co.uk/history/historic_figures/snow_john.shtml

9. *Douglas. D. (1966).* Purity and Danger. An analysis of concept of pollution and taboo. p 45

10. *Douglas. D. (1966).* Purity and Danger. An analysis of concept of pollution and taboo. p 45

11. *Douglas. D. (1966).* Purity and Danger. An analysis of concept of pollution and taboo. p 45

12. *Strasser. S. (1999)* Waste and Want. Page 200

13. *Wikipedia. (2017).* Wilhelm Röntgen. Retrieved from URL: https://en.wikipedia.org/wiki/Wilhelm_R%C3%B6ntgen

14. *Wikipedia. (2017).* Apollo 8. Retrieved from URL: https://en.wikipedia.org/wiki/Apollo_8

15. *Wikipedia. (2017).* Saturn V. Retrieved from URL: https://en.wikipedia.org/wiki/Saturn_V#Cost

16. *NASA. (1968).* Earthrise. Retrieved from URL: http://www.nasa.gov/multimedia/imagegallery/image_feature_1249.html

17. *NASA. (1968).* Earthrise. Retrieved from URL: http://www.nasa.gov/multimedia/imagegallery/image_feature_1249.html

18. *Wikipedia. (2017).* Earth rise. Retrieved from URL: https://en.wikipedia.org/wiki/Earthrise

19. *Wikipedia. (2017).* Archibald MacLeish. Retrieved from URL: https://en.wikipedia.org/wiki/Archibald_MacLeish

20. *Riley. C. (2012).* Apollo 40 years on: how the moon missions changed the world for ever. Retrieved from URL: https://www.theguardian.com/science/2012/dec/16/apollo-legacy-moon-space-riley

21. *Wikipedia. (2017).* History of climate change science. Retrieved from URL: https://en.wikipedia.org/wiki/History_of_climate_change_science

22. *Climate Action Tracker. (2016).* Effect of current pledges and policies on global temperature. Retrieved from URL: http://climateactiontracker.org/global.html

Ends.

Chapter 3 cont...

23. *Ramped. C. (2008)*. Economix. Adam Smith Would Not Approve. Retrieved from URL: http://economix.blogs.nytimes.com/2008/10/15/adam-smith-would-not-approve/?_r=0

24. *Economist. (2013)*. Crash Course. Retrieved from URL: http://www.economist.com/news/schoolsbrief/21584534-effects-financial-crisis-are-still-being-felt-five-years-article

25. *Wikipedia. (2017)*. E-democracy. Retrieved from URL: https://en.wikipedia.org/wiki/E-democracy

26. *Cisco. (2016)*. The Zettabyte Era — Trends and Analysis. Retrieved from URL: http://www.cisco.com/c/en/us/solutions/collateral/service-provider/visual-networking-index-vni/vni-hyperconnectivity-wp.html

27. *Wikipedia. (2017)*. Zettabyte. Retrieved from URL: https://en.wikipedia.org/wiki/Zettabyte

Chapter 4

1. *Allen. R. C. (2001)*. The Great Divergence in European Wages and Prices from the Middle Ages to the First World War. Retrieved from URL: http://www.nuffield.ox.ac.uk/users/allen/greatdiv.pdf

2. *Wikipedia. (2017)*. Department Store. Retrieved from URL: https://en.wikipedia.org/wiki/Department_store

3. *Seaton. A. V. (1986)* Cope's and the Promotion of Tobacco in Victorian England. European Journal of Marketing. pp 5-26.

4. *Wikipedia. (2017)*. History of advertising in Britain. Retrieved from URL: https://en.wikipedia.org/wiki/History_of_advertising_in_Britain

5. *Wikipedia (2017)*. Thorstein Veblen. Retrieved from URL: https://en.wikipedia.org/wiki/Thorstein_Veblen#Conspicuous_consumption

6. *Veblen. T. (1957)*. The Theory of the Leisure Class, Veblen. pp.73-74

7. *Strasser. S. (1999)*. Waste and Want. p. 24

8. *Strasser. S. (1999)*. Waste and Want. p. 24

9. *Strasser. S. (1999)*. Waste and Want. p. 24

10. *Wikipedia. (2017)*. Christine Frederick. Retrieved from URL: https://en.wikipedia.org/wiki/Christine_Frederick

11. *Strasser. S. (1999)*. Waste and Want. p. 197

12. *Frederick. C. (1923)*. Selling Mrs. Consumer. p 249

13. *Frederick. C. (1923)*. Selling Mrs. Consumer. p 249

14. *Strasser. S. (1999)*. Waste and Want. p. 275

15. *Wikipedia. (2017)*. Phoebus Cartel. https://en.wikipedia.org/wiki/Phoebus_cartel

16. *Automotive News. (2008)*. Annual model change was the result of affluence, technology, advertising. Retrieved from url: http://www.autonews.com/article/20080914/OEM02/309149950/annual-model-change-was-the-result-of-affluence-technology-advertising

17. *Strasser. S. (1999)*. Waste and Want. p. 275

18. *Automotive News. (2008)*. Annual model change was the result of affluence, technology, advertising. Retrieved from url: http://www.autonews.com/article/20080914/OEM02/309149950/annual-model-change-was-the-result-of-affluence-technology-advertising

19. *Automotive News. (2008)*. Annual model change was the result of affluence, technology, advertising. Retrieved from url: http://www.autonews.com/article/20080914/OEM02/309149950/annual-model-change-was-the-result-of-affluence-technology-advertising

20. *Ryan. A. Trumbull. G. Tufano. P. (2010)*. A Brief Postwar History of US Consumer Finance. Harvard Business School. Retrieved from URL: http://www.hbs.edu/faculty/Publication%20Files/11-058.pdf

21. *Woolsey. B. and Starbuck Gerson. (2016)*. The history of credit cards. Retrieved from URL: http://www.creditcards.com/credit-card-news/credit-cards-history-1264.php

22. *Klaffke. P. (2003)*. Spree: A Cultural History of Shopping. p 22.

23. *Wikipedia. (2017)*. Mail Order. Retrieved from URL: https://en.wikipedia.org/wiki/Mail_order

24. *Wikipedia. (2017)*. Home Shopping. Retrieved from URL:https://en.wikipedia.org/wiki/Home_shopping

25. *Nasdaq. (2017)*. Home Shopping Network. Common stock price. Retrieved from URL: http://www.nasdaq.com/symbol/hsni

26. *World Wide Web Foundation. (2015)*. History of the Web. Retrieved from URL:http://webfoundation.org/about/vision/history-of-the-web/

27. *World Wide Web Foundation. (2015)*. History of the Web. Retrieved from URL:http://webfoundation.org/about/vision/history-of-the-web/

28. *Zaroban. S. (2016).* US. e-commerce grows 14.6% in 2015. Retrieved from URL: https://www.internetretailer.com/2016/02/17/us-e-commerce-grows-146-2015

29. *Alderman. L. (2015).* In Sweden, a cash free future nears. Retrieved from URL: http://www.nytimes.com/2015/12/27/business/international/in-sweden-a-cash-free-future-nears.html?_r=0

Chapter 5

1. *Wikipedia. (2017).* Sensory processing. Retrieved from URL: https://en.wikipedia.org/wiki/Sensory_processing

2. *Wikipedia (2017).* Episodic Memory. Retrieved from URL: https://en.wikipedia.org/wiki/Episodic_memory

3. *Suddendorf. T. Corvallis M.C. (2007).* The evolution of foresight: What is mental time travel, and is it unique to humans? Retrieved from URL: http://www.ncbi.nlm.nih.gov/pubmed/17963565

4. *Tulving. E. (2002).* Episodic Memory: From Mind to Brain. Annual Review of Psychology. Retrieved from URL: http://www.annualreviews.org/doi/abs/10.1146/annurev.psych.53.100901.135114

5. *Pelham. B. (2004).* Affective Forecasting: The Perils of Predicting Future Feelings. American Psychological Association. Retrieved from URL: http://www.apa.org/science/about/psa/2004/04/pelham.aspx

6. *McKenna. P. (2005)* I can make you thin.

7. *Wikipedia. (2017).* Death Drive. Retrieved from URL: https://en.wikipedia.org/wiki/Death_drive

8. *Becker. E (1973)* The denial of death. Terror management theory.

9. *Kaiser. T. Kanner. A. (2004).* Psychology and Consumer Culture. American Psychological Association. Retrieved from URL: http://ase.tufts.edu/gdae/CS/Lethal%20Consumption.pdf

10. *Kaheman. D. (2011)* Thinking fast and slow. p 381.

11. *Kaheman. D. (2011)* Thinking fast and slow. p 379.

12. *Christie. S. (2015).* The easiest way to cancel Sky TV. The Telegraph. Retrieved from URL: http://www.telegraph.co.uk/finance/personalfinance/household-bills/11549715/The-easiest-way-to-cancel-Sky-TV.html

13. *Fuchs Ebaugh. H. R. (1988).* Becoming an Ex. The Process of Role Exit.

14. *Wikipedia. (2017).* Libor Scandal. Retrieved from URL:https://en.wikipedia.org/wiki/Libor_scandal

15. *Clark. D. (2011).* A complete guide to carbon offsetting. The Guardian. Retrieved from URL: http://www.theguardian.com/environment/2011/sep/16/carbon-offset-projects-carbon-emissions

Chapter 6.

1. *Neupert . R. (1995)* The End. Narration and closure in cinema. p 35

2. *MacArthur. E. J. (1990)* Extravagant Narratives: Closure and Dynamics in the Epistolary Form. p 16.

3. *Lanouette J. (2012).* A History of Three-Act Structure. Screen talks. Retrieved from URL: https://www.screentakes.com/an-evolutionary-study-of-the-three-act-structure-model-in-drama/

4. *Wikipedia. (2017)* Horace. Retrieved from URL: https://en.wikipedia.org/wiki/Horace

5. *Wikipedia (2017)* Gustav Freytag. Retrieved from URL: https://en.wikipedia.org/wiki/Gustav_Freytag

6. *Wikipedia (2017)* Dramatic Structure. Retrieved from URL: https://en.wikipedia.org/wiki/Dramatic_structure

7. *Herrstein Smith. B (1974)* Poetic Closure.

Chicago University Press.

8. *Neupert . R. (1995)* The End. Narration and closure in cinema. p 21

9. *Genette. G. (1980)* Narrative Discourse: an essay in method. p 29.

10. *Larsen. (2014)* Creating the right brand voice. Retrieved from URL: http://larsen.com/insights/creating-the-right-brand-voice/

11. *Murray N, Grierson T, Fear D, Collins S. T. (2015)* Best: 'Friday Night Lights'. Retrieved from URL: http://www.rollingstone.com/tv/lists/end-game-tvs-best-and-worst-series-finales-20150512/best-friday-night-lights-20150512

12. *Rouse. R and Abernathy T. (2014)* Death to the 3 act structure. Retrieved from URL: http://www.gdcvault.com/play/1020050/Death-to-the-Three-Act

Chapter 6 cont...

13. *Rouse. R and Abernathy T.* (2014) Death to the 3 act structure. Retrieved from URL: http://www.gdcvault.com/play/1020050/Death-to-the-Three-Act

14. *Tank J.* (2016) Playing for time. Retrieved from URL: http://www.wired.com/2016/01/that-dragon-cancer/

15. *Tank J.* (2016) Playing for time. Retrieved from URL: http://www.wired.com/2016/01/that-dragon-cancer/

16. *Tank J.* (2016) Playing for time. Retrieved from URL: http://www.wired.com/2016/01/that-dragon-cancer/

Chapter 7.

1. *Office for National Statistics.* (2016). Divorces in England and Wales. Retrieved from URL: https://www.ons.gov.uk/peoplepopulationandcommunity/birthsdeathsandmarriages/divorce/datasets/divorcesinenglandandwales

2. *Wikipedia.* (2017) Marriage. Retrieved from URL: https://en.wikipedia.org/wiki/Marriage#cite_note-297

3. *Wikipedia.* (2017) Marriage. Retrieved from URL: https://en.wikipedia.org/wiki/Marriage#cite_note-297

4. *Wikipedia.* (2017) Marriage. Retrieved from URL: https://en.wikipedia.org/wiki/Marriage#cite_note-297

5. *Wikipedia.* (2017) History of Marriage. Retrieved from URL: https://en.wikipedia.org/wiki/Marriage#History_of_marriage

6. *BBC.* (2013). How many Roman Catholics are there in the world? Retrieved from URL: http://www.bbc.com/news/world-21443313

7. *Pope John Paul ii, Vatican library (1997).* Catechism of the Catholic Church - The sacrament of Matrimony. Retrieved from URL: http://www.vatican.va/archive/ccc_css/archive/catechism/p2s2c3a7.htm

8. *Jayaram V.* (2016) Hindu Marriage, Past and Present. Retrieved from URL: http://www.hinduwebsite.com/hinduism/h_marriage.asp

9. *Pope John Paul ii, Vatican library (1997).* Catechism of the Catholic Church - The sacrament of Matrimony. Retrieved from URL: http://www.vatican.va/archive/ccc_css/archive/catechism/p2s2c3a7.htm

10. *Pope John Paul ii, Vatican library (1997).* Catechism of the Catholic Church - The sacrament of Matrimony. Retrieved from URL: http://www.vatican.va/archive/ccc_css/archive/catechism/p2s2c3a7.htm

11. *Doshi. V.* (2016). India grants divorce to man whose wife refused to live with in-laws. The Guardian. Retrieved from URL: https://www.theguardian.com/world/2016/oct/08/india-divorce-man-cruelty-western-thought-in-laws

12. *Safi M.* (2016) 'Talaq' and the battle to ban the three words that grant India's Muslim men instant divorce. The Guardian. Retrieved from URL: https://www.theguardian.com/world/2016/oct/20/talaq-and-the-battle-to-ban-the-three-words-that-grant-indias-muslim-men-instant-divorce?CMP=share_btn_link

13. *Safi M.* (2016) 'Talaq' and the battle to ban the three words that grant India's Muslim men instant divorce. The Guardian. Retrieved from URL: https://www.theguardian.com/world/2016/oct/20/talaq-and-the-battle-to-ban-the-three-words-that-grant-indias-muslim-men-instant-divorce?CMP=share_btn_link

14. *Ansari A.* (2013). Buried Alive | Clip: Marriage is an Insane Proposal | Netflix Retrieved from URL: https://www.youtube.com/watch?v=cYdsWtku9gg

15. *British Government* (2017) Get a divorce. Retrieved from URL: https://www.gov.uk/divorce/overview

16. *eHarmony* (2017) About eHarmony. Retrieved from URL:http://www.eharmony.com/about/eharmony/

17. *eHarmony* (2016). the 10 biggest reasons people fall out of love. Retrieved from URL: http://www.eharmony.com/dating-advice/relationships/the-10-biggest-reasons-people-fall-out-of-love/

18. *Wikipedia* (2017) no-fault divorce. Retrieved from URL: https://en.wikipedia.org/wiki/No-fault_divorce

19. *National Bureau of Economic Research.* (2003). Bargaining in the Shadow of the Law: Divorce Laws and Family Distress. Retrieved from URL: http://www.nber.org/papers/w10175

20. *Silverman* (2015) Divorced Couples Are Taking Awesome Selfies Together To Mark The Occasion. Buzzed. Retrieved from URL: https://www.buzzfeed.com/craigsilverman/divorced-couples-are-taking-awesome-selfies-together-to-mark?utm_term=.oyJ4LgxkE#.ifMJ8YyIV

Ends.

Chapter 8

1. *Wikipedia.* (2017) Service (economics). Retrieved from URL: https://en.wikipedia.org/wiki/Service_(economics)

2. *U.S. Department of Commerce Economics and Statistics Administration Office of Policy Development* (1996) Service Industries and economic performance. Retrieved from URL: http://www.esa.doc.gov/sites/default/files/serviceindustries_0.pdf

3. *Reference for business (2016).* Service Industry. Retrieved from URL: http://www.referenceforbusiness.com/management/Sc-Str/Service-Industry.html

4. *Deloitte (2016)* 2016 Banking industry disruptors: Banking reimagined. Retrieved from URL: https://www2.deloitte.com/us/en/pages/financial-services/articles/banking-industry-disrupter.html

5. *Carney J. (2009).*Chase Mortgage Ad From 2005 Is Funny And Scary. Retrieved from URL: http://www.businessinsider.com/chase-mortgage-ad-from-2005-is-funny-and-scary-2009-6?IR=T

6. *Treanor J. (2016)* PPI claims - all you need to know about the mis-selling scandal. Retrieved from URL: https://www.theguardian.com/business/2016/aug/02/ppi-claims-all-you-need-to-know-about-the-mis-selling-scandal

7. *Parliament.uk (2012-13)* Panel on mis-selling and cross-selling. Retrieved from URL: http://www.publications.parliament.uk/pa/jt201213/jtselect/jtpcbs/writev/misselling/sj015.htm

8. *Marthon Vik. P. (2014).* The Demise of the Bank Branch Manager. The Depersonalisation and Disembedding of Modern British Banking. Retrieved from URL: http://usir.salford.ac.uk/32045/3/PhD_thesis_amended_(1).pdf p 97

9. *Marthon Vik. P. (2014).* The Demise of the Bank Branch Manager. The Depersonalisation and Disembedding of Modern British Banking. Retrieved from URL: http://usir.salford.ac.uk/32045/3/PhD_thesis_amended_(1).pdf

10. *Edmonds T. (2013)* The Independent Commission on Banking: The Vickers Report & the Parliamentary Commission on banking standards. Retrieved from URL: http://researchbriefings.parliament.uk/ResearchBriefing/Summary/SN06171

11. *Yunus. M. (1999)* The banker to the poor. p 108

12. *El Issa. E. (2016)* 2016 American Household Credit Card Debt Study. Retrieved from URL: https://www.nerdwallet.com/blog/credit-card-data/average-credit-card-debt-household/

13. *Ofcom (2013).* Ofcom publishes research on payday loan TV adverts. Retrieved from URL: http://media.ofcom.org.uk/news/2013/Ofcom-research-payday-loan-TV-adverts/

14. *El Issa. E. (2016)* 2016 American Household Credit Card Debt Study. Retrieved from URL: https://www.nerdwallet.com/blog/credit-card-data/average-credit-card-debt-household/

15. *El Issa. E. (2016)* 2016 American Household Credit Card Debt Study. Retrieved from URL: https://www.nerdwallet.com/blog/credit-card-data/average-credit-card-debt-household/

16. *Lamagna. M. (2016)* These Americans are the most embarrassed about their credit-card debt. Retrieved from URL: http://www.marketwatch.com/story/these-americans-are-the-most-embarrassed-about-their-credit-card-debt-2016-01-26

17. *Insurance Information Institute (2015)* Industry Overview. INSURANCE INDUSTRY AT A GLANCE. Retrieved from URL: http://www.iii.org/fact-statistic/industry-overview

18. *Marquand. B. (2016)*Unclaimed billions: Are you owed a life insurance payout?Retrieved from URL: http://www.usatoday.com/story/money/personalfinance/2016/06/11/unclaimed-life-insurance-money-payout/85718732/

19. *Marquand. B. (2016)*Unclaimed billions: Are you owed a life insurance payout?Retrieved from URL: http://www.usatoday.com/story/money/personalfinance/2016/06/11/unclaimed-life-insurance-money-payout/85718732/

20. *NAUPA. (2017).* What is Unclaimed Property. Retrieved from URL: https://www.unclaimed.org/what/

21. *PT Direct (2016)* Attendance, Adherence, Drop out and Retention. Retrieved from URL: http://www.ptdirect.com/training-design/exercise-behaviour-and-adherence/attendance-adherence-drop-out-and-retention-patterns-of-gym-members

22. *Which? (2013)* How to cancel your gym membership. Retrieved from URL: http://www.which.co.uk/consumer-rights/advice/how-to-cancel-your-gym-membership

23. *Federal Trade Commission (2012)* Joining a gym. Retrieved from URL: https://www.consumer.ftc.gov/articles/0232-joining-gym

Ends.

Chapter 8 cont...

24. *Gov.uk* (2013) Health and fitness clubs: unfair contract terms. Retrieved from URL: https://www.gov.uk/cma-cases/health-and-fitness-clubs-unfair-contract-terms

25. *Evans. R.* (2015) How much will I need in retirement – and how much do I need to save now? The Daily Telegraph. Retrieved from URL:http://www.telegraph.co.uk/finance/personalfinance/special-reports/11416356/How-much-will-I-need-in-retirement-and-how-much-do-I-need-to-save-now.html

26. *Social Security Administration* (2010) Retirement & Survivors Benefits: Life Expectancy Calculator. Retrieved from URL: https://www.ssa.gov/cgi-bin/longevity.cgi

27. *Read. S.* (2013) A quarter of adults have lost a pension pot, says survey. Retrieved from URL: http://www.independent.co.uk/money/pensions/a-quarter-of-adults-have-lost-a-pension-pot-says-survey-8567984.html

28. *Norton-Taylor R.* (2016) Replacing Trident will cost at least £205bn, campaigners say. Retrieved from URL: https://www.theguardian.com/uk-news/2016/may/12/replacing-trident-will-cost-at-least-205-billion-campaign-for-nuclear-disarmament

29. *Wikipedia* (2017) Schrödinger's cat. Retrieved from URL: https://en.wikipedia.org/wiki/Schr%C3%B6dinger%27s_cat

30. *Wikipedia* (2017) Bernard Brodie. Retrieved from URL: https://en.wikipedia.org/wiki/Bernard_Brodie_(military_strategist)

Chapter 9

1. *Mintel.* (2017, April, 20) Retrieved from URL: http://www.mintel.com/global-new-products-database

2. *The International Organisation of Motor Vehicle Manufacturers* (2016) Retrieved from URL: http://www.oica.net/

3. *The International Organisation of Motor Vehicle Manufacturers* (2016) Retrieved from URL: http://www.oica.net/category/sales-statistics/

4. *Statista* (2017) Outlook Report: Consumer Electronics. Trends, Insights & Top Players. Retrieved from URL: https://www.statista.com/outlook/251/100/consumer-electronics/worldwide#takeaway

5. *Patterson. D.* (1996). A Companion to Philosophy of Law and Legal Theory. p 9

6. *Organisation For Economic Co-Operation And Development* (2001) The Control Of Transboundary Movements Of Wastes Destined For Recovery Operations.Retrieved from URL: https://www.oecd.org/env/waste/42262259.pdf

7. *E.Pongracz,V.J.Pohjola* (2004) Re-defining waste, the concept of ownership and the role of waste management. p144. Retrieved from URL: https://www.researchgate.net/profile/Eva_Pongracz/publication/222231620_Re-defining_waste_the_concept_of_ownership_and_the_role_of_waste_management/links/00b7d537f740395774000000/Re-defining-waste-the-concept-of-ownership-and-the-role-of-waste-management.pdf

8. *E.Pongracz,V.J.Pohjola* (2004) Re-defining waste, the concept of ownership and the role of waste management. p144. Retrieved from URL: https://www.researchgate.net/profile/Eva_Pongracz/publication/222231620_Re-defining_waste_the_concept_of_ownership_and_the_role_of_waste_management/links/00b7d537f740395774000000/Re-defining-waste-the-concept-of-ownership-and-the-role-of-waste-management.pdf

9. *Hoornweg D. , Bhada-Tata P. & Kennedy C.*(2013) 'Environment: Waste production must peak this century' Retrieved from URL: http://www.nature.com/news/environment-waste-production-must-peak-this-century-1.14032

10. *Braungart M. and McDonough W.* (2008) Cradle to Cradle. pp. 27-28

11. *Orange R.* (2016) Waste not want not: Sweden to give tax breaks for repairs. Retrieved from URL: https://www.theguardian.com/world/2016/sep/19/waste-not-want-not-sweden-tax-breaks-repairs?CMP=share_btn_tw

12. *Hoornweg D. , Bhada-Tata P. & Kennedy C.* (2013) 'Environment: Waste production must peak this century' Retrieved from URL: http://www.nature.com/news/environment-waste-production-must-peak-this-century-1.14032

13. *The Guardian* (2008) Waste not, want not. Retrieved from URL:http://www.theguardian.com/world/gallery/2008/aug/05/japan.recycling

14. *Ellen Macarthur Foundation* (2015) What is the circular economy? Retrieved from URL:https://www.ellenmacarthurfoundation.org/circular-economy

Ends.

Chapter 9 cont...

15. *McDonough W. and Braungart M.* (2013) The Upcycle. Forward.

16. *Kees Baldé. The United Nations University* (2015). E-waste statistics. Retrieved from URL: http://i.unu.edu/media/ias.unu.edu-en/project/2238/E-waste-Guidelines_Partnership_2015.pdf

17. *Electronics Take Back.* (2016) Facts and figures on E-waste and Recycling. Retrieved from URL: http://www.electronicstakeback.com/wp-content/uploads/Facts_and_Figures_on_EWaste_and_Recycling1.pdf

18. *Ely C.* (2014) The life expectancy of electronics. Retrieved from URL: http://www.cta.tech/Blog/Articles/2014/September/The-Life-Expectancy-of-Electronics

19. *Walton. A.* (2013) Life Expectancy of a Smartphone. Retrieved from URL: http://smallbusiness.chron.com/life-expectancy-smartphone-62979.html

20. *Kondo M.* (2010) The life changing magic of tidying.

21. *Mooallem J.* (2009) The self storage self. Retrieved from URL: http://www.nytimes.com/2009/09/06/magazine/06self-storage-t.html?em&_r=1

22. *Kondo M.* (2010) The life changing magic of tidying. p47

23. *Kondo M.* (2010) The life changing magic of tidying. p151

24. *Kondo M.* (2010) The life changing magic of tidying. p205

25. *Kondo M.* (2010) The life changing magic of tidying. p70

26. *Kretschmer A.* (2000) Mortuary Rites for Inanimate Objects The Case of Hari Kuyo Retrieved from URL: https://nirc.nanzan-u.ac.jp/nfile/2725

27. *Kretschmer A.* (2000) Mortuary Rites for Inanimate Objects The Case of Hari Kuyo. Retrieved from URL: https://nirc.nanzan-u.ac.jp/nfile/2725

28. *Kamiya S.*(2006) Last rites for the memories as beloved dolls pass away. Retrieved from URL:http://www.japantimes.co.jp/life/2006/10/15/to-be-sorted/last-rites-for-the-memories-as-beloved-dolls-pass-away/#.Vp0HJZOLSV4

Chapter 10

1. *Internet Society* (2017) A Brief History of the Internet & Related Networks. Retrieved from URL: http://www.internetsociety.org/internet/what-internet/history-internet/brief-history-internet-related-networks

2. *Kemp S.* (2016) Digital in 2016. Retrieved from URL: http://wearesocial.com/uk/special-reports/digital-in-2016

3. *Lenhart A.* (2015) Teens, Social Media & Technology Overview 2015. Retrieved from URL: http://www.pewinternet.org/2015/04/09/teens-social-media-technology-2015/

4. *Meeker M.* (2016) Internet Trends 2016. Retrieved from URL: http://www.kpcb.com/internet-trends

5. *Statistic Brain* (2016) YouTube Company Statistics Digital Technology. Retrieved from URL: http://www.statisticbrain.com/youtube-statistics/

6. *Mayer-Schonberger V.* (2011) Delete: The Virtue of Forgetting in the Digital Age. p 117.

7. *Mayer-Schonberger V.* (2011) Delete: The Virtue of Forgetting in the Digital Age. p 117.

8. *Statistic Brain* (2016) Average Cost of Hard Drive Storage. Retrieved from URL:http://www.statisticbrain.com/average-cost-of-hard-drive-storage/

9. *Martine J van Bennekom* (2015) A case of digital hoarding. British Medical Journal. http://casereports.bmj.com/content/2015/bcr-2015-210814.full?sid=19141a77-de2f-4dcd-9b98-d079622oef63

10. *Robson J.* (2015) 'Overnight, everything I loved was gone': the internet shaming of Lindsey Stone. Retrieved from URL: http://www.theguardian.com/technology/2015/feb/21/internet-shaming-lindsey-stone-jon-ronson

11. *Sherry Turkle* (2012) Connected but alone. TED Talks. Retrieved from URL: https://www.ted.com/speakers/sherry_turkle

12. *Davey G.* (2016) Social Media, Loneliness, and Anxiety in Young People. Retrieved from URL: https://www.psychologytoday.com/blog/why-we-worry/201612/social-media-loneliness-and-anxiety-in-young-people

13. *Davey G.* (2016) Social Media, Loneliness, and Anxiety in Young People. Retrieved from URL: https://www.psychologytoday.com/blog/why-we-worry/201612/social-media-loneliness-and-anxiety-in-young-people

Ends.

14. *NHS Digital by NatCen Social Research and the Department of Health Sciences, University of Leicester* (2014) Mental Health and Wellbeing in England. Retrieved from URL: http://content.digital.nhs.uk/catalogue/PUB21748/apms-2014-full-rpt.pdf

15. *Trust* (2016) 2016 TRUSTe/NCSA Consumer Privacy Infographic – US Edition. Retrieved from URL: https://www.truste.com/resources/privacy-research/ncsa-consumer-privacy-index-us/

16. *Nave K.* (2016) Infoporn: how VPN use varies by country. Retrieved from URL: http://www.wired.co.uk/article/vpn-use-worldwide-privacy-censorship

17. *Wikipedia* (2017) Revenge Porn. Retrieved from URL:https://en.wikipedia.org/wiki/Revenge_porn

18. *Lenhart A. Ybarra M. Price-Feeney M.* (2016) Nonconsensual Image Sharing: One In 25 Americans Has Been A Victim Of "Revenge Porn". Retrieved from URL:https://datasociety.net/pubs/oh/Nonconsensual_Image_Sharing_2016.pdf

19. *Lenhart A. Ybarra M. Price-Feeney M.* (2016) Nonconsensual Image Sharing: One In 25 Americans Has Been A Victim Of "Revenge Porn". Retrieved from URL: https://datasociety.net/pubs/oh/Nonconsensual_Image_Sharing_2016.pdf

20. *Unterleider N.* (2014) Who Is Liable When Cloud Services Are Hacked? Retrieved from URL: https://www.fastcompany.com/3035104/who-is-liable-when-cloud-services-are-hacked

21. *Lenhart A. Ybarra M. Price-Feeney M.* (2016) Nonconsensual Image Sharing: One In 25 Americans Has Been A Victim Of "Revenge Porn". Retrieved from URL: https://datasociety.net/pubs/oh/Nonconsensual_Image_Sharing_2016.pdf

22. *Symantec* (2015) State of Privacy Report 2015. Retrieved from URL: http://www.symantec.com/content/en/us/about/presskits/b-state-of-privacy-report-2015.pdf

23. *Apple* (2017) Apple Media Services Terms and Conditions. Retrieved from URL: https://www.apple.com/legal/internet-services/itunes/us/terms.html

24. *Hern A.* (2015) I read all the small print on the internet and it made me want to die. Retrieved from URL: http://www.theguardian.com/technology/2015/jun/15/i-read-all-the-small-print-on-the-internet

25. *Guardian* (2017) Terms of Service. Retrieved from URL: http://www.theguardian.com/help/terms-of-service

26. *Wikipedia* (2017) Right to be Forgotten. Retrieved from URL: https://en.wikipedia.org/wiki/Right_to_be_forgotten

27. *Wikipedia* (2017) Rehabilitation of Offenders Act 1974. Retrieved from URL: https://en.wikipedia.org/wiki/Rehabilitation_of_Offenders_Act_1974

28. *European Commission* (2016) Fact sheet on the "right to be forgotten" ruling. Retrieved from URL: http://ec.europa.eu/justice/data-protection/files/factsheets/factsheet_data_protection_en.pdf

29. *European Commission* (2016) Fact sheet on the "right to be forgotten" ruling. Retrieved from URL: http://ec.europa.eu/justice/data-protection/files/factsheets/factsheet_data_protection_en.pdf

30. *Constine J.* (2016) Facebook Climbs To 1.59 Billion Users And Crushes Q4 Estimates With $5.8B Revenue. Retrieved from URL: https://techcrunch.com/2016/01/27/facebook-earnings-q4-2015/

31. *Rowe A.* (2016) WhatsApp, Facebook Messenger Growth Outpaces Instagram, Twitter. Retrieved from URL: http://tech.co/messaging-apps-growth-outpaces-instagram-twitter-2016-06

32. *Morrison K.* (2016) Is Snapchat Growing Faster Than Instagram? (Infographic). Retrieved from URL: http://www.adweek.com/socialtimes/is-snapchat-growing-faster-than-instagram-infographic/641668

33. *Bitly* (2016) the top 20 social media trends in 2016, according to the experts. Retrieved from URL: https://bitly.com/blog/the-top-20-social-media-trends-in-2016-according/

34. *eMarketer* (2016) Snapchat to Grow 27% This Year, Surpassing Rivals. Retrieved from URL: http://www.emarketer.com/Article/Snapchat-Grow-27-This-Year-Surpassing-Rivals/1014058#sthash.YuRlQ4h5.dpuf

35. *eMarketer* (2016) Snapchat to Grow 27% This Year, Surpassing Rivals. Retrieved from URL: http://www.emarketer.com/Article/Snapchat-Grow-27-This-Year-Surpassing-Rivals/1014058#sthash.YuRlQ4h5.dpuf

Ends.

Chapter 11

1. *Eirlys Roberts (1966)* The Consumers. p.17
2. *Competition & Markets Authority (2016)* Energy Market Investigation. Retrieved from URL: https://assets.publishing.service.gov.uk/media/576c23e4ed915d622c000087/Energy-final-report-summary.pdf
3. *Competition & Markets Authority (2016)* Energy Market Investigation. Retrieved from URL: https://assets.publishing.service.gov.uk/media/576c23e4ed915d622c000087/Energy-final-report-summary.pdf
4. *Gov.uk (2013)* Bank account switching service set to launch. Retrieved from URL: https://www.gov.uk/government/news/bank-account-switching-service-set-to-launch
5. *Hyde D. (2011)* Easy seven-day bank account switching to arrive in UK 'from 2013'. Retrieved from URL: http://www.thisismoney.co.uk/money/saving/article-2036543/Current-account-switching-Easy-seven-day-bank-switching-UK-2013.html
6. *Gov.uk (2013)* Bank account switching service set to launch. Retrieved from URL: https://www.gov.uk/government/news/bank-account-switching-service-set-to-launch
7. *Gov.uk (2013)* Bank account switching service set to launch. Retrieved from URL: https://www.gov.uk/government/news/bank-account-switching-service-set-to-launch
8. *Gov.uk (2013)* Bank account switching service set to launch. Retrieved from URL: https://www.gov.uk/government/news/bank-account-switching-service-set-to-launch
9. *Energy switch Guarantee (2013)* Switching with confidence. Retrieved from URL: https://www.energyswitchguarantee.com/
10. *BBC (2017)* Energy bills: customer switching hits six year high. Retrieved from URL: http://www.bbc.co.uk/news/business-39095434
11. *AOL. Money. (2017)* 1m customers switch bank accounts but rush slows. Retrieved from URL: http://money.aol.co.uk/2017/01/24/1m-customers-switch-bank-accounts-but-rush-slows/
12. *Failcon (2017)* Failcon goes global. Retrieved from URL: http://thefailcon.com/index.html
13. *CB Insights (2017)* 204 Startup Failure Post-Mortems. Retrieved from URL: https://www.cbinsights.com/blog/startup-failure-post-mortem/
14. *Carroll. R. (2014)* Silicon Valley's culture of failure ... and 'the walking dead' it leaves behind. Retrieved from URL: https://www.theguardian.com/technology/2014/jun/28/silicon-valley-startup-failure-culture-success-myth
15. *Apple (2016)* App Store Improvements. Retrieved from URL: https://developer.apple.com/news/?id=09012016a
16. *Perez S. (2016)* Apple's big App Store purge is now underway. Retrieved from URL: https://techcrunch.com/2016/11/15/apples-big-app-store-purge-is-now-underway/
17. *Wikipedia (2017)* Kia Cee'd. Retrieved from URL: https://en.wikipedia.org/wiki/Kia_Cee%27d
18. *Foxall J. (2014)* Kia's seven-year warranty success. Retrieved from URL: http://www.telegraph.co.uk/motoring/car-manufacturers/kia/10683878/Kias-seven-year-warranty-success.html
19. *Costello M. (2016)* Kia credits seven-year warranty for sales boom. Retrieved from URL: http://www.caradvice.com.au/502707/kia-credits-seven-year-warranty-for-sales-boom/
20. *Foxall J. (2014)* Kia's seven-year warranty success. Retrieved from URL: http://www.telegraph.co.uk/motoring/car-manufacturers/kia/10683878/Kias-seven-year-warranty-success.html
21. *Drucker P. F. (2006)* Classic Drucker from the pages of the Harvard Business Review. Page 26-27
22. *Drucker P. F. (2006)* Classic Drucker from the pages of the Harvard Business Review. Page 28
23. *Drucker P. F. (2006)* Classic Drucker from the pages of the Harvard Business Review. p. 29
24. *Statistic (2017)* Nokia's net profit/loss* from 2006 to 2016 (in million euros). Retrieved from URL: https://www.statista.com/statistics/267820/nokias-net-income-since-2006/
25. *Drucker P. F. (2006)* Classic Drucker from the pages of the Harvard Business Review. p. 29
26. *Smith P. (2013)* The Nokia insider who knows why it failed warns Apple it could be next. Retrieved from URL: http://www.afr.com/technology/the-nokia-insider-who-knows-why-it-failed-warns-apple-it-could-be-next-20130906-jh3iz
27. *Wikipedia (2017)* Supermarkets. Retrieved from URL: https://en.wikipedia.org/wiki/Supermarket
28. *Economist (2010)* You Choose. Retrieved from URL: http://www.economist.com/node/17723028
29. *Economist (2010)* You Choose. Retrieved from URL: http://www.economist.com/node/17723028
30. *Vizard S. (2017)* How Aldi became Britain's fifth largest supermarket. Retrieved from URL: https://www.marketingweek.com/2017/02/07/aldi-fifth-largest-supermarket/

Ends.

Chapter 11 cont...

31. *Ruddick G. (2014)* It may already be too late for Tesco and Sainsbury's, the rise of Aldi and Lidl looks unstoppable. Retrieved from URL: http://www.telegraph.co.uk/finance/newsbysector/retailandconsumer/10974773/It-may-already-be-too-late-for-Tesco-and-Sainsburys-the-rise-of-Aldi-and-Lidl-looks-unstoppable.html

32. *YouGov Brand Index (2016)* 2016 Annual Rankings: UK. Retrieved from URL: http://www.brandindex.com/ranking/uk/2016-annual

33. *Ruddick G. (2014)* It may already be too late for Tesco and Sainsbury's, the rise of Aldi and Lidl looks unstoppable. Retrieved from URL: http://www.telegraph.co.uk/finance/newsbysector/retailandconsumer/10974773/It-may-already-be-too-late-for-Tesco-and-Sainsburys-the-rise-of-Aldi-and-Lidl-looks-unstoppable.html

34. *Iyengar S. (2017)* Business Insider. Retrieved from URL: http://www.businessinsider.com/too-many-choices-are-bad-for-business-2012-12?op=1&r=US&IR=T&IR=T/#iminate-choices-to-make-decision-making-easier-in-your-business-13

35. *Phil Gyford (2017)* Our Incredible Journey. Retrieved from URL: https://ourincrediblejourney.tumblr.com/

36. *Phil Gyford (2017)* Our Incredible Journey. Retrieved from URL: https://ourincrediblejourney.tumblr.com/

37. *Cooney S. (2012)* Adam Smith, Milton Friedman and the Social Responsibility of Business. Retrieved from URL: http://www.triplepundit.com/2012/08/adam-smith-milton-friedman-social-responsibility-of-business/

38. *Cooney S. (2012)* Adam Smith, Milton Friedman and the Social Responsibility of Business. Retrieved from URL: http://www.triplepundit.com/2012/08/adam-smith-milton-friedman-social-responsibility-of-business/

39. *Becker-Olsen, K. L. Cudmore B. A.; Hill R. P. (2006)* "The impact of perceived corporate social responsibility on consumer behaviour." Journal of Business Research.

40. *Kardashian K. (2013)* When Retailers Do Good, Are Consumers More Loyal? Retrieved from URL: http://www.tuck.dartmouth.edu/news/articles/when-retailers-do-good-are-consumers-more-loyal

41. *Collinson P. (2014)* Do's and don'ts of getting rid of your old car. The Guardian. http://www.theguardian.com/money/2014/feb/14/getting-rid-old-car-scrap

42. *Gov.uk (2017)* Driver and Vehicle. Retrieved from URL: https://www.gov.uk/government/organisations/driver-and-vehicle-licensing-agency

43. *Collinson P. (2014)* Do's and don'ts of getting rid of your old car. The Guardian. http://www.theguardian.com/money/2014/feb/14/getting-rid-old-car-scrap

Conclusion

1. *Nuland S. B. (1993).* How We Die. p. 16

2. *Kermode F. (2000).* A Sense of an Ending. p. 4

Index

A

Abstinence 46
Actionable by the user 238
Adam Smith 66
Adjust 216
Advanced Study of
Sustainability 171
advertisers 26
Advertising 100, 111, 210
adverts 100
Age Concern 23, 159
agricultural 53
Air Travel 227
Alan Segal 33
Aldi 221
Aldus Manutius 85
Alkmeon 239
Amazon 86
Amazon Prime Air 87
American Express 84
AMEX 84
ancestors 30
Angelika Kretschmer 179
Animism 179
Anna Bertha 61
Anti-amnesia 188
Apollo 8 62
App 20
Apple 98, 174

Apple's iCloud 196
Apple's iTunes 197
App Store 216
Archibald MacLeish 63
Aristotle 119, 239
Arlington Cemetery 191
Aziz Ansari 139

B

Babylonians 32
Banks 149
Barbara Herrstein Smith
120
Barry Schwartz 220
BBC 141
Bernard Brodie 160
Bible 35, 136
Blackberry 219
Black Sea 37
Bob Circosta 85
Bomb proof 186
Book of Data 33
Bourgogne 38
Brand Index 221
brands 22
Britain 54
Britain's National
Consumer Council 112
British Medical Journal
190

Broadwick Street 58
Bruno Bowden 215
Buddha 34
Buddhists 34
Bud Paxson 85
burials 32
Burials 31
Business 209

C

Cadillac 82
Calvinist 43
cancer 237
Canon 235
carbon 65
carbon offsetting 112
cars 164
Car scrapping 227
Catholic 38, 41, 136, 138
Sacrament of Extreme
Unction 38
Catholic Church 39
Catholicism 40
Ceremony
ceremony 30
CERN 86
Chevrolet 82
cholera 58
Christian 35, 138
Christine Frederick 78, 80

Christmas 178
Churches 36
circular economy 169
Cisco 68
climate change 104
Closure 30, 105
Closure Experiences 25
Closure in the Novel 122
colonoscopy 107
Computer Fraud and
Abuse Act 196
Conference of Parties 64
Consciously connected
235
consumer 54, 60, 67, 84
consumer electronics 164
consumer lifecycle 19, 50
Consumer relationship 21
consumers 18, 50, 77
Consumers 25, 51, 73
Cope Bros & Co 76
Corporate Social
Responsibility 225
Cradle to Cradle 167
Credit cards 152, 153
credit crisis 147
Credit Crunch 67
Currency 88
customer lifecycle 23, 25

Ends.

Customer lifecycle 122

D

Daniel Kaheman 106
Data Policy Facebook 200
Data Protection 109
Data storage 189
dead 30
Dead
deceased 32
death 41
Death 30, 39, 45
dying 32
Death instinct 102
Death Master File 156
Deborah Hendersen 127
Defining the moment of
waste 165
Delete 188
Deloitte 147
Denial of Death 103
dénouement 120
Department for Work and
Pensions 159
Department of Work and
Pensions 23
Dexter 123
dieting 101
Digital 18, 23, 184
Diners Club 83
Disney 40
divorce 135
Divorce 138, 141
DNR 238
Do Not Resuscitate 238
Dropbox 191
dump 167

E

Earth 36
Earth Day 63
Earthrise 62, 63
eat 102
Effective forecasting 99
Egg 154
Egyptian 33
E-Harmony 141
Eirlys Roberts 73
Elizabeth MacArthur 118
Ellen Macarthur
Foundation 169
Emotional Triggers 236
Endel Tulving 97
endings 27

End of Life Vehicle
Directive 228
Energy Ombudsman 213
England 53
environmental issues 225
Ernest Becker 103
Europe 74
European Union 204
E-waste 170
experiencing self 107
Extravagant Narratives 118

F

Facebook 138, 191, 199
Facebook Messenger 205
factories 54
FailCon 215
fail-fast 222
Fair Repair 173
fasting 46
film 118
Financial Ombudsman
148
Financial Reserve 66
financial services 23
Florida Office of Insurance
Regulation 156
Ford model-T 81
four classes of waste 166
France 38
Frank Borman 62
Frank McNamara 83
Frank Nuovo 219
Freud 102
Friday Night Lights 123
Friends of the Earth 63
Funeral for needles 179

G

Galen Rowell 63
Game 124
Game over 124
Games 130
General Electric 81
General Motors 82
Genoese 37
George Monbiot 112
Gerard Genette 122
Germs 61
Gilbert and Wilson 99
Givry 38
Global warming 17
Gods
Attim 33
God 34

Osiris 33
Goodbyes 180
Good Housekeeping 78
Google 202
Gordon B 37
Grameen Bank 151
graves 30, 31
greenhouse gases 64
Grim Reaper 39
Gustav Freytag 120
Gyms 157

H

hard drives 190
Harding, Howell & Co's 75
Harvard Business Review
219
Heaven 33, 36
Heavens 35
Helen Rose Fuchs Ebaugh
109
Hello's 180
Heretics 38
Hillary Clinton 68
Hindu 36, 137
Hindu Marriage Act 138
Hoarders 174
Home shopping 85
Home Shopping Network
85
Horace 119
humans 30

I

IBM 98
Ilona Cheyne 166
Independent Commission
on Banking 150
Industrial Revolution 75
Instagram 205
insurance 211
internet 85
iPad 187
Islamic 35, 138
Quran 35
Islamists 36
ISP 184
Israel 31
iZettle 88

J

Jacob de Geer 88
James Lovell 62
Janet Gunter 170
Japan 168

Jasper in Indiana 60
Jean-Jacques Rousseau 50
Jehovah Witnesses 34, 36
Jenn Frank 130
Jews 38
Johannes Gutenberg 42
John Biggins 83
John Calvin 44
John Everett Millais 76
John Kenneth Galbraith
80
John Snow 58
Jordan 31
Joseph Stiglitz 225
Julia McNair Wright 78
Jumping the Shark 123

K

Kasser and Sheldon 104
Ken Wong 124
Kia cars 216
Kick-Starter 224
kitchen 50
kitchens 88
Kleenex 60
Kruglanski and Webster
104

L

labourers 53
landfill 166
landfills 22
Latter-day Saints 37
Lauren Chakkalackal 193
Lawrence Kitson 154
Libor 111
Lidl 221
Life Changing Magic of
Tidying 174
lifecycle 17
Life insurance 156
likes 23
Lindsey Stone 191
loan 151
love-interest 119
Lydia Brasch 174

M

Macbeth 119
Marianna Torgovnick 122
Marie Kondo 174
Marie Kondo technique
177
Marketing 210
marriage 134

Ends.

Marriage 137
Martin Luther 42
Martin Mayer 81
Mary Douglas 59
Mary Meeker's Annual Internet Trends 187
Maslow's hierarchy of needs 26
Max Weber 34, 43
Meaningful goodbyes 222
Memorial for Dolls 179
memory 97
Memory 97
Mesopotamia 37
Michael Braungart 167
Michael T. Hannan 54
Microsoft 127
Middle East 31
Middle Palaeolithic 31
Milton Friedman 225
Mintel 163
Monument Valley 124
Mormons 34, 36
mortgage 21, 152
Mrs. Consumer 78
Muhammad Yunus 151

N

narrative 118
National Association of Unclaimed Property Administrators 156
Natufians 31
Nature 167
Neanderthal 30
Need For Closure Scale 105
Neolithic 32
NerdWallet 153
Netflix 195
New York Times 63
NHS Digital 192
Nicolaus Copernicus 41
Ninety-Five Theses 42
Nirvana 34
Nobel Prize 225
No-fault divorce 143
Nokia 98, 219
nuclear 61
nuclear deterrent 159
Nuffield College Oxford 74

O

Ofcom 153
off-boarding 25
Off-boarding 20, 60, 209, 214
off-site storage 175
On-boarding 21, 96, 209
on-boarding 19, 26
organisational management 54
Organisation for Economic Co-operation and Development 165
Osiris 33
ourincrediblejourney 224

P

packaged goods 22
Pal Marathon Vik 149
Paradox of Choice 220
Paul McKenna 101
Paying off interest 150
Payment Protection Insurance - PPI 148
Peak End Rule 106
Pears Soap 76
Pebble 223
pension 20
Pension 158
pensioner 23
pensions 23
Peter Drucker 218
Pew Research Centre 187
Phil Gyford 222
Philip Brickman 99
Philip Lieberman 31
Philippe Aries 45
Phoebus Cartel 81
photo 23
Pieter Bruegel's The Triumph of Death 39
Plague 37, 38, 41, 97
Plain Tobacco Packaging 237
PlayStation 4 125
Poetic Closure 120
pollution 17
Pope 42
post-visual 61
Priests 38
printer ink cartridge 235
Private property 165
Proctor & Gamble 221

producers 26
product 20, 163
Product euthanasia 80
product lifecycle 26
Products 17, 21
Professor Mayer-Schonberger 188
Professor Susan Strasser 54
Progressive Obsolescence 78
Protestantism 42, 43, 46
Pryce Pryce-Jones 85
Psychology 96
PT Direct 157

Q

Quran 36

R

Rag and Bone man 56
Readmill 224
re-cycling 217
recycling 168
Relics 40
Religion 135
Religions 58
remembering self 107
Renaissance 41
Restart Project 170
Revenge Porn 195
Richard Neupert 131
Richard Rouse 127
Right to be Forgotten 201, 203
Right To Be Forgotten 69
right to repair 172
Riot Games 127
Role Exit 109
Rolling Stone 123
Roseanne 124
Rubbish dumps 22
Russian revolution 144

S

Sacrament of Matrimony 137
Satan 39
Saturn V engine 62
Schrödinger's cat 160
Scotland 66
Scrap Metal Dealers Act 228
Seiko J. Pohjola 165
self-actualisation 25

self-harming 193
Service Design 17
services 18, 22, 146
sewage system 60
Shakespearean 119
Shannon and Chris Neuman 144
Shannon McSheffrey 136
share photos 24
Sharon Beder 44
Sherry Turkle 192
Sherwin Nuland 239
Sicilian 37
SimCity 125
Simply Signature Loan 147
Sky 108
smoking 237
Snapchat 205, 238
Social dilemmas 111
social media 187
Social Networks 227
starting experiences 19
Starting Experiences 25
Steam Review Watch 126
subprime 148
Sub-Prime' 67
sub-prime mortgages 147
Susan Strasser 57
Sweden 88, 168
switching accounts 150
Syria 31
Systematic abandonment 218

T

taxes 134
Tech Crunch 216
telephone 85
Tenzing Chusang 138
TERMINATION 198
Terms and Conditions 20, 197
Terror management theory 103
Tesco 221
That Dragon Cancer 129
The Economist 68
The End 118
The First Supper 83
The Guardian 200
Theory of the Leisure Class 77
The Price of Materialism 154

Ends.

The Upcycle 170
Thibaut and Kelly 110
Thinking Fast and Slow 106
Thomas and George Cope 76
Thomas J. Barratt 76
Thorstein Veblen 77
three act structure 119
Tim Berners Lee 86
Timely 238
Tini Owens 141
Tithing 37
Tom Abernathy 127
Too big to fail 65
trash 22
Trident 159
Tuck Business School 226
TV 100
TV advertising 153
TV adverts 121

U

Ugo Vallauri 170
UK energy watchdog Ofgem 211
Uniform Penny Post 85
usage phase 20
US Bureau of Labor Statistics 147
US Defence Advanced Research Projects Agency 186
US Department of Commerce Economics and Statistics Administration 147

V

Vatican 136
Virtual Private Networks 194
VPNs 194
VR 124

W

Waste 56, 168
waste-creation 165
Which? 149
Whole Earth Catalogue 63
Wilhelm Röntgen 61
William Anders 62
William Lucas 85
William McDonough 167
Wired 194
World War 83
World Wide Web 86

X

X-Ray 61

Z

Zombie Apps 216

'1-Click' 86
5'i's 146
5'i's of services 146
5x5 approach 234
7 Day Switch Guarantee 212

Made in the USA
Middletown, DE
07 February 2019